A COGNITIVE ANALYSIS OF U.S. AND CHINESE STUDENTS' MATHEMATICAL PERFORMANCE ON TASKS INVOLVING COMPUTATION, SIMPLE PROBLEM SOLVING, AND COMPLEX PROBLEM SOLVING

by

Jinfa Cai

NATIONAL COUNCIL OF TEACHERS OF MATHEMATICS

Copyright © 1995 by
THE NATIONAL COUNCIL OF TEACHERS OF MATHEMATICS, INC.
1906 Association Drive, Reston, Virginia 22091-1593

Library of Congress Cataloging-in-Publication Data:

Cai, Jinfa.
    A cognitive analysis of U.S. and Chinese students' mathematical
performance on tasks involving computation, simple problem solving,
and complex problem solving / by Jinfa Cai.
        p.   cm. — (Journal for research in mathematics education.
Monograph, ISSN 0883-9530 ; no. 7)
    Includes bibliographical references.
    ISBN 0-87353-424-7 (pbk.)
    1. Mathematics—Study and teaching—United States.
2. Mathematics—Study and teaching—China. 3. Mathematical ability—
Cross-cultural studies. I. Title. II. Series.
QA13.C33 1996
153.9′451′08951—dc20   95-50590
                                                              CIP

Printed in the United States of America

TABLE OF CONTENTS

ACKNOWLEDGMENTS

I would like to take this opportunity to express my deepest appreciation to the many individuals who assisted and influenced my work. I am especially grateful to the editor, Douglas Grouws, for his professional editorial assistance, sustained encouragement, and prompt processing of every aspect of this work for publication. I am also grateful to several anonymous reviewers who made valuable comments concerning an earlier version of this manuscript, thereby contributing to its improvement. Appreciation is extended to Jim Stigler and Ken Travers for the last-minute inquiries.

The study reported here is based on my doctoral dissertation, completed under the direction of Edward Silver, at the School of Education, University of Pittsburgh, in 1994. I am very grateful for his encouragement, support, and excellent intellectual guidance. Working with Ed has been most challenging and productive. I am also grateful to Gaea Leinhardt, Suzanne Lane, Mark Ginsburg, and Wilbur Deskins for their intellectual guidance and encouragement.

Appreciation is extended to the Alumni Committee of the University of Pittsburgh School of Education for providing research funds to conduct this study, to Richard Mayer for his permission to use his tasks to assess word problem solving component skills, to Harold Stevenson and James Stigler for their permission to use their tasks to assess computation skills, and to Edward Silver and Suzannne Lane for their permission to use some QUASAR open-ended problems to assess complex problem solving skills. Thanks also go to the Doctoral Association of Educators of the University of Pittsburgh for selecting me to receive the Outstanding Dissertation Award.

In addition, I would like to express my gratitude to the many individuals who helped with data collection, data analysis, or proofreading. Special thanks go to Professors Bingyi Wang and Chuanhan Lu for their assistance in collecting Chinese data, and to Patricia Kenney for her final round of proofreading. My gratitude also extends to all the students who particpated in this study and to their teachers, principals, and parents.

Special thanks go to all my teachers, past and present, as well as future, for their great influence on me. Teachers are the true heroes who make presidents, scholars, good citizens, and scientists. Teachers truly make a difference in this world.

Finally, very special thanks go to my beloved and beautiful wife, NingWang, for her love, patience, understanding, and intellectual support and assistance.

DEDICATION

I would like to dedicate this work to my mom, my very first teacher, who taught me God's love and the value of education. She also taught me to love and respect people, to work hard, and to be persistent.

ABSTRACT

The mathematical performance of 250 U.S. sixth-grade students from both private and public schools and 425 Chinese sixth graders from both key and common schools was examined on multiple-choice tasks assessing computation and simple problem solving and on open-ended tasks assessing complex problem solving. Chinese students performed significantly better than U.S. students on both computation and simple problem solving. The results were about the same for the two samples on complex problem solving. Moreover, when subsets of U.S. and Chinese students were matched on their computational performance, the U.S. students scored significantly higher than comparable Chinese students on the measures of both simple and complex problem solving.

U.S. and Chinese students had similar overall performance on complex problem solving, but a detailed cognitive analysis of students' written responses revealed not only many similarities in the solutions but also many subtle differences. For example, the types of strategies employed and the types of errors made by the Chinese students were similar to those for the U.S. students, although the Chinese students' solutions tended to be more elegant. Also, U.S. students tended to use visual representation more frequently than Chinese students, who tended to use symbolic representation (e.g., algebraic equations) more frequently.

The results of this study suggest not only the complexity of examining mathematical performance differences, but also the inadequacy of using a limited range of tasks to measure mathematical performance in cross-national studies. One of the main contributions of this study is its use of a variety of mathematical tasks to capture the thinking and reasoning of U.S. and Chinese students. Another contribution is the scheme used to analyze student performance, a scheme based not solely on the percentage correct or incorrect, but rather on a detailed analysis of students' strategies, representations, and errors. This range of tasks and the associated methodology supported the discovery of findings of similarities and differences between U.S. and Chinese students that have not been reported previously.

1. INTRODUCTION

PURPOSE OF THE STUDY

The main purpose of this study was to conduct a cognitive analysis of mathematical performance by a group of U.S. sixth-grade students and a comparable group of Chinese sixth-grade students, with particular attention to complex problem solving as well as to computation and simple word problem solving. Through the cognitive analysis, this study served as a means of examining cross-national similarities and differences in these several forms of mathematical performance by U.S. and Chinese students. In this way, the study was designed to provide more in-depth information about the thinking processes used by students in the two countries than has been available from prior cross-national studies. To achieve this purpose, U.S. and Chinese students' mathematical performance on tasks involving computation, simple problem solving, and complex problem solving was assessed.

In the last several decades, a number of researchers have conducted cross-national studies to compare the mathematical performance of students in the U.S. and in Asian countries such as China, Japan, and Korea (e.g., Becker, 1992; Cai & Silver, 1995; Husen, 1967; Lapointe, Mead, & Askew, 1992; Robitaille & Garden, 1989; Song & Ginsburg, 1987; Stevenson & Lee, 1990; Stevenson & Stigler, 1992; Stevenson, Lee, Chen, Lummis, Stigler, Liu, & Fang, 1990; Stigler, Lee, & Stevenson, 1990). Most of these researchers have used multiple-choice items to measure students' mathematical performance, and they have paid particular attention to procedural knowledge (i.e., computation) and routine applications. Other aspects of mathematical performance, such as more complex problem solving, have been less often investigated. Only recently, a few researchers have started to pay attention to complex problem solving in cross-national studies in mathematics (e.g., Becker, 1992; Cai & Silver, 1995; Ito-Hino, 1994). Hatano (1988) has argued that having expertise in routine applications does not imply expertise in complex and novel problem solving. Thus, the cross-national studies conducted to date have achieved good coverage of procedural knowledge and routine problem-solving skills; but more complex forms of mathematical performance, especially as related to complex problem solving, remain to be investigated.

In this study, a set of seven complex, open-ended mathematical problems was administered to a cross-national sample to assess students' higher-level mathematical thinking. Open-ended problem formats allow students to construct solutions and to provide a visible record of their solution processes that may include critical cognitive aspects of mathematical thinking. The use of open-ended problems allows the researcher to examine cognitive aspects of students' mathematical thinking and reasoning, such as solution strategies, mathematical misconceptions, mathematical justifications, and modes of representation. The experience of successfully using open-ended problems in the California Assessment Program

(1989) and in the QUASAR project (e.g., Cai, Lane, & Jakabcsin, in press-a; Cai, Magone, Wang, & Lane, in press-b; Lane, 1993; Lane & Silver, in press; Magone, Cai, Silver, & Wang, 1994; Silver, 1993; Silver & Cai, 1993) shows that open-ended problems are appropriate formats to assess and evoke students' higher-level mathematical thinking. Along with the set of seven complex, open-ended mathematical problems, two other types of mathematical problems were used to assess U.S. and Chinese students' computation skills and component processes of solving simple word problems.

The solving of mathematical word problems has long been an important and popular topic of study by cognitive psychologists. For example, Mayer (1987) analyzed the solving of arithmetic word problems and identified four cognitive component processes: translation, integration, planning, and execution. For the purpose of this study, the first three components (translation, integration, and planning) were distinguished from the fourth (execution) in the testing and some of the analyses. A set of 18 multiple-choice questions designed by Mayer and his associates was used to assess students' performance in the component problem-solving processes of translation, integration, and planning (hereafter, called the *component questions*). A set of 20 multiple-choice computation tasks was used to assess students' execution or computation skills (hereafter, called the *computation tasks*). Although many previous cross-national studies have used computation tasks to measure students' mathematical performance, little or no error analysis of performance has been done. This study examined not only the relative success rates, but also the errors made by U.S. and Chinese students in solving computation tasks and component questions. Taken together, these two sets of tasks (computation tasks and component questions) provide the basis for an analysis of U.S. and Chinese students' mathematical performance on various cognitive component processes of solving word problems.

This study extended earlier work by Mayer, Tajika, and Stanley (1991), who examined U.S. and Japanese fifth graders' mathematical performance by using computation tasks and component questions. Mayer et al. (1991) reported that Japanese students outperformed their U.S. counterparts on both types of problems. However, an additional analysis showed that when samples of U.S. and Japanese students were equated on the basis of their level of computational performance, U.S. students scored higher than comparable Japanese students on the component questions. This subtle difference was only visible through the use of both computation tasks and component questions. Similarly, in this study, computation tasks were used to match Chinese and U.S. students on computational performance. Then component questions were used to examine differences between U.S. and Chinese students' component problem-solving processes. In addition, U.S. and Chinese students' performance on the open-ended problems was examined and compared on the basis of matching the computational performance.

Cai and Silver (1995) proposed that the magnitude of the differences between American and Chinese students on mathematical understanding and mathematical applications might be smaller than that on procedural knowledge and symbol manipulation. The use of computation tasks, component questions, and open-ended problems in a single study provides an opportunity to test this conjecture directly.

WHY CROSS-NATIONAL STUDIES IN MATHEMATICS?

There is a long history of cross-national comparative studies in education (Brickman, 1988; Medrich & Griffith, 1992; Postlethwaite, 1988). Comparative studies not only provide information on students' achievement examined in the context of the world's varied educational institutions, but also help to identify effective aspects of educational practice. Postlethwaite (1988) identified four objectives of comparative studies:

- Identifying what is happening in different countries that might help improve educational systems and outcomes
- Describing similarities and differences in educational phenomena between systems of education and interpreting why these exist
- Estimating the relative effects of variables that are thought to be determinants of educational outcomes (both within and between systems of education)
- Identifying general principles concerning educational effects

These four objectives have been implemented in many comparative studies, especially in the designs of the International Association for the Evaluation of Educational Achievement (IEA) studies (Medrich & Griffith, 1992).

To advise the National Center for Education Statistics and the National Science Foundation on U.S. participation in international comparative studies of education, the Board on International Comparative Studies in Education presented a conceptual framework for study that identified several reasons why the U.S. should participate in international studies (Bradburn & Gilford, 1990):

- Improving understanding of education systems
- Providing information on the students' achievement in relation to the much broader range of the world's education systems
- Identifying the factors that do and do not promote educational achievement
- Enhancing the research enterprise itself
- Recording the diversity of educational practice
- Promoting issue-centered studies

There is a great deal of consistency between the objectives of comparative studies identified in Bradburn and Gilford (1990) and in Postlethwaite (1988). The reasons for doing comparative studies, as posited by Bradburn and Gilford (1990) and summarized below, may be suitable for any nation and can be read as such by substituting the nation's name for every reference to the United States.

International comparative research on education provides an important addition to research within the United States. It increases the range of experience necessary to improve the measurement of educational achievement; it enhances confidence in the generalizability of studies that explain the factors important in educational achievement; it increases the probability of the dissemination of new ideas to improve the design or management of schools and classrooms; and it increases the research capacity of the United States as well as that of other countries. Finally, it provides an opportunity to chronicle practices and policies worthy of note in their own right. (1990, p. 4)

The significance of doing cross-national studies in mathematics is related not only to the objectives of comparative educational studies, but also to the nature, importance, and usefulness of mathematics. Many people believe that because "numbers are numbers" and "geometric figures are geometric figures," the basic number operations and the geometric figural relationships should function the same across cultures (Stigler & Perry, 1988). The generality of mathematics would seem to imply that it is one of the school subjects least affected by culture.[1] Stigler and Perry stated that "it is the relatively transcultural nature of mathematics that makes it especially interesting for cross-cultural study" (1988, p.195). In strongly culture-related learning content areas, such as history, not only does the context of learning vary across cultures, but so does the content of what is being learned. However, in mathematics, the content remains similar even though the cultures vary. Therefore, cross-national studies in mathematics are an appropriate approach for educators, sociologists, and psychologists to understand the major aspects of the educational systems in different cultures.

The importance and usefulness of mathematics have been widely articulated by the international mathematics education community (e.g., Cao & Cai, 1989; Cockcroft, 1982; Fujita, Miwa, & Becker, 1990; Grouws, 1992; National Council of Teachers of Mathematics [NCTM], 1989; State Education Commission of China [SECC], 1987). In particular, Robitaille and Garden indicated that

> mathematics is seen as contributing to the intellectual development of individual students, as preparing them to live as informed and functioning citizens in contemporary society, and as providing students with the potential to take their places in the fields of commerce, industry, technology, and science. (1989, p. 2)

More importantly, mathematics is viewed no longer as just a prerequisite subject but rather as a fundamental aspect of literacy for a citizen in contemporary society (Mathematical Sciences Education Board [MSEB], 1993; NCTM, 1989). In view of the importance of mathematics for society and for individual students, the efficacy of mathematics teaching and learning in schools deserves sustained scrutiny. Cross-national studies in the domain of mathematics provide mathematics educators with opportunities to identify effective ways of teaching and learning mathematics in a wider cultural context. Examination of what is happening in the learning of mathematics in other societies helps researchers and educators to understand how mathematics is taught by teachers and is learned and performed by students in different cultures. It also helps them to

---

[1]It is important to note, however, that not everyone agrees with this. In fact, some people argue that mathematics is a culturally-bound subject (e.g., Ernest, 1991).

reflect on theories and practices of teaching and learning mathematics in their own culture.

In particular, examination of cognitive similarities and differences in mathematical performance in different cultures could help identify students' current status in a developmental model of mathematical thinking and reasoning, and provide diagnostic and decision-making information. The information gathered from cognitive-based cross-national studies may indicate particular successes or difficulties of students' mathematics learning and performance and provide important information to those responsible for their learning as they try to decide how to resolve students' learning difficulties.

EXAMINING COGNITIVE ASPECTS OF CROSS-NATIONAL PERFORMANCE DIFFERENCES: RATIONALE AND FEASIBILITY

In the past several decades, great progress has been made in examining patterns of mathematical performance and in understanding the factors contributing to performance differences in mathematics between students in the U.S and in some Asian countries. In particular, previous cross-national studies have provided a large body of knowledge about students' mathematical achievement in different cultures and about which cultural and educational factors might influence their learning of mathematics. However, almost all previous cross-national studies have used a limited range of tasks to measure narrow mathematical performance and have reported the quantitative performance differences (e.g., differences in mean scores) rather than a qualitative analysis of the differences. Separate analyses in terms of symbol manipulation and mathematical understanding have rarely been reported. Few analyses of strategies and errors in students' mathematical problem solving across cultures exist (Hatano, 1990). A detailed picture of how students differ in their mathematical learning and cognition across cultures would be more informative than a list of mean scores for a broad spectrum of groups, including classroom teachers, parents, educational policy makers, educational and psychological researchers, politicians, and public media (Robitaille, 1992).

There are a few exceptions in which researchers (e.g., Becker, 1992; Cai & Silver, 1994, 1995; Ito-Hino, 1994; Mayer et al., 1991; Silver et al., 1995) have made attempts to examine cross-national differences and similarities in mathematics by focusing on the cognitive aspects of mathematical problem solving. For example, Cai and Silver (1995) examined U.S. and Chinese students' cognitive difficulties in solving a story problem that involved division with a remainder. They found that Chinese students outperformed U.S. students in the computational phase of solving the problem, but that students in both samples had similar cognitive difficulties in the "sense-making" or interpretation phase of solving the problem.

As mentioned earlier, Mayer et al. (1991) examined the performance of U.S. and Japanese fifth graders on computation and component questions. Overall, Japanese fifth graders outperformed their counterparts on both types of problems, but when U.S. and Japanese students were matched on the basis of their level of computational performance, U.S. students scored higher than Japanese students on the component questions of measuring problem-solving skills involving translation, integration, and planning. In another study, Silver et al. (1995) analyzed the responses of Japanese and U.S. students to a task requiring multiple solutions of a single problem, in which a picture of a set of marbles arranged in a certain way was given to students and they were asked to determine the number of marbles in as many ways as they could. They found that Japanese students performed better than U.S. students with respect to the proportions of correct solutions; however, many subtle differences and similarities in the solution strategies employed and the modes of explanation of the solution processes were noted. For example, Japanese students used mathematical symbolic representations in their solutions more frequently than the U.S. students.

Although these few studies have limitations, such as the use of a limited number of tasks to examine students' cognitive aspects of mathematical problem solving, they suggest the significance of cognitive-based cross-national studies. They also suggest the feasibility of such investigations. This study examined cross-national similarities and differences in mathematical performance by conducting a cognitive analysis of student responses to a wide array of tasks. The focus draws from the recognized importance of problem solving in mathematics education (e.g., Charles & Silver, 1988; NCTM, 1989; Silver, 1987; Schoenfeld, 1985) and in cognitive psychology (e.g., Simon, 1979, 1989).

Problem solving is an important part of the school mathematics curriculum. Studies of mathematical problem solving have already moved from a focus only on the product (i.e., the actual solution or answer) to a focus on the process (i.e., the set of planning and executing activities that direct the solution search). As Robitaille and Travers (1992) have argued, in a cross-national study, the "[i]nformation about how students approach the solution of a given problem is .... more important than whether or not they are able to recognize the correct solution" (p. 708). A cross-national comparison based on the cognitive aspects of mathematical problem solving will assist the transition from comparing learning products to comparing learning processes across national boundaries.

Cognitive theorists view the outcome of learning as a developing cognitive skill. Recent advances in cognitive psychology provide background information about how people construct their knowledge and highlight the importance of examining more than the correctness of the answer to a problem (Royer, Cisero, & Carlo, 1993). The large body of research in cognitive psychology provides a basis for cross-national comparisons of the cognitive aspects of students' problem-solving processes. During the last several decades, great progress has been made in understanding learning and problem-solving mechanisms, knowledge structure, and the acquisition of domain expertise. For example, cognitive scientists studying domain expertise indicate that experts and novices differ not only in the capability of solving a problem, but they also seem to process information in qualitatively different ways, with experts using more adequate representations and solution strategies and having fewer misconceptions (Chi, Glaser, & Farr, 1988; Simon, 1979, 1989).

Researchers have suggested that cognitive theories and research methods may apply to educational assessments (e.g., Frederiksen, Glaser, Lesgold, & Shafto, 1990; Freedle, 1990; Glaser, 1987; Ronning, Glover, Conoley, & Witt, 1985; Snow & Lohman, 1989; Wittrock & Baker, 1991). The advances of cognitive psychology contribute not only to the identification of important cognitive aspects of measuring students' performance (Glaser, Lesgold, and Lajoie, 1985; Royer et al., 1993) but also to the actual ways in which students' performance is measured (Lesgold, Lajoie, Logon, & Eggan, 1990). Moreover, well-constructed measures of student cognitive processes can provide useful information for classroom teachers to understand learners, to diagnose learning difficulties, and to improve instruction in schools (Wittrock, 1990).

In cognitive research, psychologists have made heavy use of techniques such as collecting and analyzing "think aloud" protocols (Ericsson & Simon, 1980, 1984). Verbal protocol methods are powerful and adequate to gain information about unfolding cognitive processes because subjects' internal states are probed using concurrent verbalization methods. However, they are not feasible for cross-national studies with a large sample because the process of collecting, coding, and analyzing verbal protocol data is extremely labor intensive. Thus, there is a need for the development of feasible assessment procedures that can examine the cognitive aspects of performance differences across samples in different nations.

Fortunately, recent progress in the development of performance assessments in mathematics has demonstrated the usefulness of open-ended problems to reveal the processes of mathematical thinking (Cai et al., in press-a; Cai et al., in press-b; California State Department of Education, 1989; Kulm, 1994; Lane, 1993; Magone, Cai, Silver, & Wang, 1994; Silver & Cai, 1993). Through analyses of a set of open-ended mathematical tasks and student responses to these tasks, Magone et al. (1994) indicated that well-developed, open-ended mathematical problems are valid and feasible for assessing students' critical cognitive aspects of mathematical problem solving.

This study provided a detailed cognitive analysis of U.S. and Chinese students' solutions to open-ended, complex mathematical problems by focusing on the following aspects: solution strategies, mathematical misconceptions, mathematical justifications, and modes of representation. This study also examined and compared cognitive component processes of U.S. and Chinese students' word problem-solving processes: that is, translation, planning, integration, and execution.

STATEMENT OF THE PROBLEM AND RESEARCH QUESTIONS

In this study, the researcher conducted a cognitive analysis of the mathematical performance of a group of U.S. sixth-grade students and a comparable group of Chinese sixth-grade students, with particular attention given to computation, simple problem solving, and complex problem solving. The study tried to provide more in-depth information about the thinking processes used by the students in the two countries than has been available from prior cross-national studies. The study also tried to demonstrate the feasibility of integrating cognitive psychology and performance assessment techniques into cross-national studies in mathematics. In particular, this study was designed to answer the following research questions:

1. What are the similarities and differences in mathematical performance between a group of U.S. students and a comparable group of Chinese students as measured by three types of tasks: performing computation tasks, assessing the component processes (translation, integration, and planning) in word problem-solving, and solving a set of open-ended problems?

2. What is the nature of the cognitive similarities and differences between a group of U.S. students and a comparable group of Chinese students in their performance in solving a set of open-ended problems, specifically: their use of strategies to solve a set of open-ended problems, their use of modes of representation, the errors they make, and the quality of the mathematical justifications they give?

3. When levels of computational performance are matched, how does a subgroup of U.S. students differ from a comparable subgroup of Chinese students in their success in answering questions assessing word-problem-solving component processes and solving a set of open-ended problems?

2. THEORETICAL BASIS OF THE STUDY

CROSS-NATIONAL STUDIES IN MATHEMATICS

In early 1964, the International Association for the Evaluation of Educational Achievement (IEA) launched the First International Mathematics Study (FIMS) (Husen, 1967). The aims of the study were to examine students' achievement in various mathematics topics in different countries and to explore some of the underlying factors contributing to notable differences in achievement (Husen, 1967; Robitaille & Garden, 1989; Travers & Westbury, 1989). The FIMS revealed that American students did not perform as well overall as Chinese (in Hong Kong), Japanese, and Korean students. Since then, a number of researchers have conducted cross-national studies of mathematics to examine performance differences and to uncover underlying factors that may lead to the varied performance of American and Asian students (e.g., Becker, 1992; Cai & Silver, 1995; Harnisch, Walberg, Tsai, Sato, & Fryans, 1985; Hess et al., 1986; Husen, 1967; Lapointe, et al., 1992; Lapointe, Mead, & Phillips, 1989; McKnight et al., 1987; Miura, 1987; Robitaille & Garden, 1989; Stevenson & Lee, 1990; Stevenson, et al., 1990; Stigler, Lee, & Stevenson, 1987; 1990; Song & Ginsburg, 1987; Theisen, Achola, & Boakari, 1983; Westbery, Ethington, Sosniak, & Baker, 1994). Two key questions that drove cross-national studies were (1) Do Asian students really outperform American students in mathematics? (2) If so, why?

Figure 1 provides a framework that summarizes previous cross-national studies in mathematics. In cross-national studies, mathematical performance has been assessed on the basis of a set of tasks. It is assumed that mathematical abilities and knowledge can also be assessed on the basis of this set of tasks (Baranes, Perry, & Stigler, 1989). Different mathematical tests have been constructed and adapted to examine the mathematical performance of students among different countries in these previous cross-national studies (Lapointe et al., 1992; Robitaille & Garden, 1989; Stevenson & Lee, 1990). After the examinations of mathematical performance, researchers tried to interpret the observed performance differences from different perspectives, such as schooling, cultural and social factors, student characteristics, and the methodological aspects of testing and sampling. For example, some researchers argued that the observed performance differences among students in different countries might be attributed to variations in mathematical curricula (McKnight et al., 1987; Westbury, 1992). Other researchers asserted that the observed performance differences might be due to sampling errors (Bracey, 1992, 1993; Rotberg, 1990).

Differences in Mathematical Performance

American and Asian students' performance differences in mathematics have been investigated across school levels and mathematical content topics in previous cross-national studies. This section contains the reviews of findings with

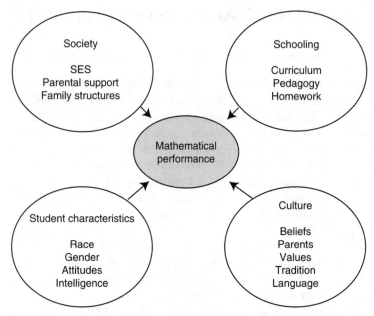

Figure 1. A framework of cross-national studies.

respect to mathematical performance differences among the students in different countries. The findings of the performance differences were summarized with a focus on U.S. students' mathematical performance compared to that of Asian students. The summary of the findings was presented in three sections: a) differences across school levels, b) differences across mathematical topic areas, and c) differences in cognitive processes.

Differences across school levels. Previous cross-national studies have compared preschool children and in-school students' performance in mathematics. This approach provided evidence of how schooling itself might contribute to performance differences. Examples of such cross-national studies at the preschool level include the studies conducted by Ginsburg and his associates (Ginsburg et al., 1990; Song & Ginsburg, 1987). Their analyses of early mathematical thinking suggested that U.S. preschool children performed as well as or even better than their counterparts in some of the Asian countries.

Song and Ginsburg (1987) extensively studied the development of informal mathematics (activities that do not involve written symbolism) and formal mathematics (activities involving the written, symbolic mathematics taught in school) in Korean and U.S. children at ages 4 through 8. Their data showed that the performance levels of U.S. preschool children at ages 4 and 5 were significantly higher than those of Korean children on a test of early mathematical knowledge. However, Korean children's performance increased consistently after the age of 5, and after the first few years of school, Korean children showed superior performance.

Song and Ginsburg (1987) concluded that the superior performance of Korean children in early mathematical thinking seemed to stem from efficient schooling.

In another study, Ginsburg et al. (1990) further compared the development of early mathematical thinking among American, Chinese, Japanese, and Korean 4-year-old preschool children without formal instruction, and Chinese preschool children with formal instruction. Chinese preschools with formal instruction have two 30-minute daily lessons in a variety of subjects. Mathematics is taught twice a week and children are assigned homework. Chinese preschools without instruction do not have these formalized classes; rather, the children's time is filled with game activities. Teachers themselves in Chinese preschools, with and without instruction, also vary in terms of their education. In the preschools with formal instruction, teachers have graduated from college or a teacher training school, whereas teachers in the no-instruction preschools are usually only high school graduates. In this study, Ginsburg et al. (1990) did not find American preschool children's performance to be superior in early mathematical thinking as noted in their earlier study (Song & Ginsburg, 1987); instead, they found that there were no significant differences among American, Chinese, Japanese, and Korean preschool children without formal instruction. However, Chinese preschool children with formal instruction outperformed American, Chinese, Japanese, and Korean preschool children without formal instruction. The schooling these Chinese preschool children were receiving might have contributed to their superior performance in early mathematical thinking.

As noted, school students have been compared across elementary, middle, and high school levels in previous cross-national studies. Examples of cross-national studies at the elementary school level are those directed by Stevenson, Stigler, and their associates (e.g., Chen & Stevenson, 1989; Stevenson & Lee, 1990; Stevenson, Stigler, & Lee, 1986; Stevenson et al., 1990; Stevenson, Stigler, Lee, Lucker, Kitamura, & Hsu, 1985; Stigler, Lee, & Stevenson, 1986; Stigler, Lee, & Stevenson, 1990). Their first study involved both kindergarten and elementary school children (Stevenson et al., 1986), whereas later they focused their attention on elementary school children: first, third, and fifth graders. In a longitudinal study, they again examined the first graders when they were in the fifth and eleventh grades (Chen & Stevenson, 1989; Stevenson & Lee, 1990; Stevenson, Chen, & Lee, 1993). These studies tried to identify reasons for the superior academic achievement of Chinese and Japanese children as compared to American children. During the past decade, a set of studies has been completed that specifically examined the correlates of children's academic achievement in two American cities, Minneapolis and Chicago; two Chinese cities, Beijing and Taipei; and one Japanese city, Sendai.

A major finding with respect to the performance difference from the studies by Stevenson was that in overall performance U.S. students consistently did not perform as well as Chinese and Japanese students. For example, in one study, Stevenson and Lee (1990) tested American, Chinese, and Japanese first- and fifth-grade students' performance on word problems and items requiring only calculation. They found that American students performed significantly lower than Chinese and Japanese students. Chinese and Japanese students had higher mean scores than their American counterparts on both types of items and at both grade levels. To more closely examine performance differences, Stevenson and Lee (1990) also analyzed the distribution of the top and bottom 100 scores in mathematics by country. They found that of those students receiving the 100 lowest scores worldwide at the first grade level, 56 were American children, and of those receiving the 100 lowest scores at the fifth grade level, 67 were American children. Only 14 American children were among those students receiving the top 100 scores at the first-grade level, and only one American child was among those receiving the top 100 scores at the fifth-grade level.

Their longitudinal studies of the first graders when they were at the fifth- and eleventh-grade levels showed similar performance differences in mathematics; that is, American students did not do as well as Asian students (Stevenson, Chen, & Lee, 1993; Stevenson & Stigler, 1992). Stevenson and his associates focused on cultural context and schooling to interpret these performance differences between American and Asian students. These findings are summarized in the next section.

Some researchers (e.g., Fuson & Kwon, 1991; Miura & Okamoto, 1989) examined U.S. and Asian elementary school students' representations of numbers and their understanding of place value and solving addition and subtraction problems. The purpose of their studies was to examine the argument that the performance differences in mathematics between U.S. and Asian students may be due, in part, to fundamental variations in the representations of numbers. The results from Miura and Okamoto (1989) showed that Japanese students have significantly greater understanding of the place-value concept than U.S. students and that this may be due to the easier representation of numbers in the Japanese language than in English. Similarly, Fuson & Kwon (1991) found that American students made more errors than Korean students in place-value tasks and multi-digit addition and subtraction because of American students' lack of linguistic support for understanding the place value of tens and ones in English.

The IEA studies compared the mathematical performance of the eighth- and twelfth-grade students to examine student achievement in various mathematics topics in different countries and to explore underlying causes for notable differences in achievement (Robitaille & Garden, 1989; Travers & Westbury, 1989). The studies have been carried out twice, the FIMS in 1964, and the Second International Mathematics Study (SIMS) in 1980. The results of the SIMS showed that Japanese students outperformed U.S. students in overall mathematics achievement at both the eighth- and twelfth-grade levels (Robitaille & Garden, 1989; Robitaille & Travers, 1992).

In addition to the IEA studies, Harnisch, Walberg, Tsai, Sato, and Fryans (1985) conducted a large comparative study between a sample of high school students (15-, 16-, and 17-year-olds) in the state of Illinois and in Japan. Their purpose was to compare mathematical performance and investigate the background that might contribute to the performance difference. Like other researchers, Harnisch et al. (1985) concluded that U.S. students did not do as well as Japanese students. Furthermore, their study confirmed a strong link between courses completed and performance.

In late 1983, the *Dallas Times-Herald* did an international survey in order to compare the educational achievement of 12-year-old students in Dallas with students from Australia, Canada, England, France, Japan, Sweden, and Switzerland (Robitaille & Travers, 1992). The results showed that among these countries, the United States scored lowest in mathematics, with a mean score of 25% correct compared to 50% (the highest mean) for Japan.

The International Assessment of Educational Progress (IAEP) has conducted two international surveys on mathematics (Lapointe et al., 1992; Lapointe et al., 1989). In the first survey, the IAEP surveyed 13-year-old students in 12 populations, including the United States, European countries, and French and English Canada. Korea was the only Asian country included in this study. On a selected set of National Educational Assessment Progress (NEAP) items scored according to a system with a mean of 500, the Korean students had the highest average mathematics proficiency score (568), and American students had the lowest score (458) (Lapointe et al., 1989). The second IAEP study, which surveyed 13-year-old students from the same 20 different countries and 9-year-old students from 14 of the 20 countries, was conducted in 1990. Asian students included in the survey were from Mainland China, Korea, and Taiwan. Results from the second study again showed that 9- and 13-year-old American students did not perform as well as their Asian counterparts.

Recently, U.S. and Japanese mathematics educators conducted a cross-national study on the problem-solving performance of students at grades 4, 6, 8, and 11 in different geographic locations in both countries. The purpose of the study was to examine American and Japanese students' problem-solving behaviors when they were asked to generate multiple solutions to a set of problems (Becker, 1992). Overall, Japanese students were able to solve the problems in more ways and to use more sophisticated solution strategies than their American counterparts. Nevertheless, there were similarities between cognitive aspects of Japanese and U.S. students' solutions (Becker, 1992). For example, both American and Japanese students used multiple solution strategies to solve a single mathematical problem.

Differences across mathematical topic areas. In order to have a valid measure of students' mathematical performance in cross-national studies, mathematics as the construct domain should be fully represented (Messick, 1989). Thus, not only should varied levels of students be studied, but also a range of representative mathematical topics should be covered in the assessment.

Some comparisons of American and Asian students' performance in mathematics have used different topics in different studies, whereas other studies have used a broad sample of representative mathematical topics. For example, the second IAEP surveyed students' performance on five topic areas[2]: numbers and operations; measurement; geometry; data analysis, statistics, and probability; and algebra and functions (Lapointe et al., 1992; Lapointe et al., 1989). Results show that Asian students performed better than American students on each of the topics. Stevenson et al. (1990) used samples of first- and fifth-grade students from Chicago (U.S.) and Beijing (China) to compare their achievement on mathematical topic areas including word problems, number concepts, mathematical operations, measurement and scaling, graphs and tables, spatial relations, visualization, estimation, and speed tests. They found that there were almost no areas in which the children in Chicago performed as well as the children in Beijing. Stigler et al. (1990) reported a similar finding to that reported in Stevenson et al. (1990) when they examined American, Japanese, and Chinese first and fifth graders' mathematical knowledge in different areas, such as computation, word problems, number concepts and equations, estimation, operations, geometry, visualization and mental folding, mental calculation, and oral problems.

The SIMS measured eighth-grade students' abilities to solve arithmetic, algebra, geometry, statistics, and measurement problems and measured twelfth-grade students' abilities to solve algebra, geometry, elementary functions and calculus, probability and statistics, sets and relations, and number-system problems. Among the participating countries, the students in the United States performed at or below the median level in each of the topic areas tested, with students from Japan and Hong Kong outperforming the U.S. students. In particular, the performance of U.S. students in arithmetic was below the average for all participating countries (Robitaille & Garden, 1989).

Some cross-national studies have focused on specific mathematical topics in examining performance differences. For example, Miura and Okamoto (1993) investigated American and Japanese students' performance in solving arithmetic word problems. They found that, overall, Japanese students performed better than their U.S. counterparts on 14 different types of arithmetic word problems.

Differences in cognitive processes. A few studies have initiated an examination of performance differences in mathematics by focusing on cognitive process levels. For example, the second IAEP study classified items into three process levels: conceptual understanding, procedural knowledge, and problem solving (Lapointe et al., 1992). The items were assigned to the three process levels on the basis of the judgment of NAEP experts. The problem-solving items were intended to assess students' higher-level thinking, whereas the procedural-knowledge items were intended to assess students' lower-level thinking, such as recall, recognition of mathematical symbols, or routine operations. The IAEP reported that Chinese and Korean students performed better than American students on all three process levels.

---

[2]The first surveys of IAEP also included a sixth topic, which was Logic and Problem Solving. Questions in this topic assessed an understanding of the tools of mathematics itself, those processes that are central to the extension and development of mathematics and its use.

However, there are two concerns with respect to the classification of items according to process levels. The first is related to the extent of agreement in the experts' judgments. It was found that the agreement between NAEP experts' judgments and other experts' judgments for assigning 1990 NAEP items to the three process levels was low. Silver, Kenney, and Salmon-Cox (1992) selected the set of 137 NAEP items at grade 8 from the 1990 mathematics assessment and asked eight experts to classify these items into the three process categories. They found that the agreement between the NAEP experts' classification and the other experts' classification was only .69. In particular, only half of the items that were classified as conceptual understanding by the NAEP experts were classified as conceptual understanding by the experts in the Silver et al. (1992) study. The second concern is related to the match between the expert classification of the process levels and the actual processes used by students in solving problems. It is difficult to determine exactly what processes students with differing backgrounds use in trying to solve problems. In fact, these categories are not intended for such use. In spite of the limitations of the item classification in the IAEP study, the idea of focusing on different levels of thinking to examine cross-national differences might be fruitful for future cross-national studies.

A few cross-national studies in mathematics have conducted a fairly detailed cognitive analyses to examine performance differences. For example, Cai and Silver (1993) examined American and Chinese sixth-grade students' cognitive difficulties in solving a story problem involving division with a remainder (DWR): "An army bus holds 36 soldiers. If 1128 soldiers are being bused to their training site, how many buses are needed?" The DWR problem is interesting because in solving this problem a student not only needs to correctly apply and carry out division but also needs to make sense of the result in the given context. The finding from the Cai and Silver (1993) study showed that Chinese sixth-grade students markedly outperformed their American counterparts with respect to performing appropriate computations. In fact, over 90% of all students (91% for U.S. students and 90% for Chinese students) selected the appropriate computations in solving the story problem. However, a larger percentage of Chinese students (83%) than American students (51%) executed the computations successfully. Chinese students did not outperform American students with respect to their sense-making of their computational results in the given context, although a larger percentage of Chinese sixth-grade students (19%) than American sixth-grade students (10%) provided the appropriate interpretations of the computational results based on the given situation.

As was mentioned in Chapter 1, Mayer et al. (1991) reported that Japanese students outperformed their American counterparts in a test of basic, computation-based mathematics achievement and a test of arithmetic word- problem-solving skills (problem translation, problem integration, and problem planning). But when American and Japanese students were matched on the basis of their level of computational performance, U.S. students scored higher than Japanese students on the test of problem solving. American and Japanese students were partitioned into five achievement levels according to students' computational performance. Then, Mayer et al. (1991) plotted the problem-solving performance for American and Japanese students for each achievement level. With respect to the three components of problem solving (problem translation, problem integration, and problem planning), they found that U. S. students performed better than Japanese students on problem integration and problem planning at all five achievement levels and on problem translation at four of the five achievement levels. However, their classification of computational achievement levels was arbitrary. In fact, the main criterion for partitioning groups was that each group have at least five students. The highest achievement level had only 5 out of a total of 132 American students, but 77 out of a total of 110 Japanese students. Although matching on computational achievement may hold promise in research such as this, the categories must be more rigorously and thoughtfully applied.

Summary of differences in mathematical performance. A general finding across almost all existing cross-national studies in mathematics was that students in Asian countries consistently outperformed U.S. students at the elementary, middle, and high school levels but not at the preschool level. This suggests that schooling might be an important factor that contributes to the performance differences in mathematics between American and Asian students. With respect to mathematical topics and related areas (e.g., computation, word problems, geometry), there are virtually no areas in which American students did as well as Asian students.

Although results from the few existing studies that examined performance differences based on cognitive aspects favor Asian students, interesting performance patterns have been noted. There are some similarities between American and Asian students in their use of solution strategies and representations and in their difficulties in solving mathematical problems. Available evidence suggests that when American students are matched with Japanese students on computation, then they are also similar in their problem-solving behavior. However, because only a few studies have focused on cognitive aspects in examining the performance differences between American and Asian students, it is too early to draw any firm conclusions. But existing studies suggest the need for, and usefulness of, a focus on cognitive analyses in conducting cross-national comparisons.

Factors Contributing to the Performance Differences

Given that Asian school students generally outperformed American students in mathematics, a number of cross-national studies have attempted to determine contributing factors. The differences have been interpreted from different perspectives, including intelligence (Lynn, 1982, 1988), educational systems (e.g., Duke, 1986), schooling (e.g., Chen & Stevenson, 1990; Robitaille & Garden, 1989), cultural contexts (e.g., Cogan, Torney-Purta, & Anderson, 1988; Hess & Azuma, 1991; Stevenson & Lee, 1990), and representation of numbers (e.g., Fuson & Kwon, 1991; Miura, 1987; Miura & Okamoto, 1989). A review of selected factors follows.

Intelligence. Because a positive relationship between children's intelligence and their academic achievement has frequently been found (Sternberg, 1982), an obvious hypothesis is that Asian children are simply smarter than American children and that the performance differences in mathematics reflect levels of intelligence. The research by Lynn (1982, 1988) seems to favor this hypothesis. On the basis of an analysis of cognitive functions, Lynn concluded that the average IQ of Japanese children exceeds that of average American children. However, Lynn's research has been shown to be problematic on the grounds that the samples from the two countries were not similar in terms of urban/rural residence and the socioeconomic background of students (Stevenson & Azuma, 1983). Stevenson et al. (1985) claimed that when compatibility in background was ensured and the children were tested with culturally appropriate materials, there were no significant differences in general level of intelligence between Japanese, Chinese, and American children. They concluded that the differences in mathematics achievement cannot be explained by intellectual differences. This conclusion seems plausible, especially in light of the fact that there were no performance differences in mathematics among American and Asian preschool children (Song & Ginsburg, 1987). If the hypothesis that Asian children are smarter than American children were true, then the Asian preschool children should have outperformed U.S. preschool children.

There are many controversies, however, regarding the definition of intelligence (Weinberg, 1989). The measurement of intelligence is even more problematic for doing cross-national intelligence testing (Segall, Dasen, Berry, & Poortinga, 1990). The culture relevance of intelligence makes it difficult to test. In fact, people in different cultures have different views of the characteristics of persons who are more, or less, intelligent and the personality attributes associated with an intelligent person (Keats, 1982). For example, a sample of Chinese people responded that an intelligent person should have the following personality attributes: perseverance, effort, determination, and social responsibility. On the other hand, Australians emphasized confidence, happiness, and effectiveness in social relations. Thus, it seems very difficult to ensure the comparability in background in testing students' intelligence from different cultures. It is equally difficult to ensure that children are tested with culturally appropriate measures. In the light of current evidence, the conclusion that the differences in mathematics achievement between American and Chinese students can be explained by intellectual differences is not plausible, and any further cross-national testing of intelligence should be done very carefully.

Early informal mathematics. Research has shown that mathematical thinking begins well before children enter school (Ginsburg, 1983). Thus, one could hypothesize that Asian children's superior performance in mathematics is probably related to an early advantage in "informal" mathematics education. This has been shown not to be the case (Ginsburg et al. 1990; Song & Ginsburg, 1987). As mentioned earlier, Song and Ginsburg (1987) found that the performance levels of U.S. children at ages 4 and 5 on all subskills of informal mathematics were significantly higher than those of Korean children. But after the first few years of school, Korean children's disadvantage disappeared rapidly. By age 7 or 8, there was a reversal, with the Korean children having superior performance in both informal and formal mathematics. Moreover, when Ginsburg et al. (1990) compared the development of early mathematical thinking among American, Chinese, Japanese, and Korean preschool children without formal instruction and Chinese preschool children with formal instruction, they found that there were no significant differences among American, Chinese, Japanese, and Korean preschool children without formal instruction. It was only the Chinese preschool children with formal instruction that outperformed American, Chinese, Japanese, and Korean preschool children without formal instruction. Therefore, Asian children do not begin with an intellectual head start. Instead, schooling is likely a factor contributing to the academic performance differences among Asian and American school students.

Representation of numbers. There is a growing number of studies examining linguistic influences on learning mathematics (e.g., Cocking & Mestre, 1991; Orr, 1987). For language minority students, language can be a barrier to success in mathematics (Orr, 1987). Thus, one may argue that the performance differences in mathematics between U.S. and Asian students may be due, in part, to the variations in the representations of numbers. Because of the congruous nature of the Asian language systems and number systems, Asian students may have an advantage in learning mathematics. Findings from some studies (e.g., Fuson & Kwon, 1991; Miura, 1987; Miura & Okamoto, 1989) seem to support this argument.

In Chinese-based languages (Chinese, Japanese, and Korean), numerical names are congruent with the traditional base 10 numeration system. In this system, the value of a given digit in a multidigit numeral depends on the face value of the digit (0–9) and on its position. The spoken numerals in Chinese and Japanese correspond exactly to their written form. For example, 11, 12, and 20 are spoken as "ten-one," "ten-two," and "two-ten" in Chinese or Japanese. In other words, the spoken name of a written number reveals the place value of that number. In English, however, the spoken names do not always conform to their written form. The names for the numerals 11, 12, and 20, spoken in English, lack the elements of tens and ones contained in them.

Research findings suggest that the above-stated irregularities of the number system in the English language have serious consequences that can adversely affect children's numerical learning in several different ways (Fuson & Kwon, 1991; Miura, 1987; Miura & Okamoto, 1989). In particular, the irregularities of number systems in the English language lead to difficulties for English-speaking children when compared to children speaking Chinese or a language based on Chinese. These difficulties for the English-speaking children arise in learning number-word sequence, adding and subtracting numbers with a sum between 10 and 18, learning adequate representations for multidigit numbers, adding and subtracting multidigit numbers accurately and meaningfully, and understanding place value (Fuson & Kwon, 1991, 1992).

Miura and Okamoto (1989) indicated that the differences in numerical languages could lead to variations in the cognitive representation of numbers. After comparing U.S. and Japanese first graders' respective representations of numbers and understanding of place value, these researchers indicated that U.S. and Japanese first graders' cognitive representations of number were different. They suggested that this difference might enhance the Japanese children's understanding of place value and their subsequent mathematical performance.

There are other linguistic differences that may contribute to performance differences in mathematics. For example, in numbering objects, a unit is always used in Chinese and Japanese. Different units are used to number different objects. For example, "36 buses" will be written or spoken as "36 LIANG buses." The LIANG is the unit for number of buses. When numbering 25 desks, the unit is ZHANG, "25 ZHANG desks." The number is hardly used apart from the unit indicating what type of thing is being numbered. In English, associating a unit with the number of objects is rare. Thus, English-speaking children may be more likely to "lose the unit."

In naming and speaking dates, there are also differences between the Chinese and English languages. In particular, in Chinese, spoken names of days and months correspond exactly to their numerical sequence. For example, Monday in spoken Chinese is "week one" and January is "one month." Similarly, in naming the number of weeks and months, a unit is also attached to the number. For example, five weeks is spoken as "five GE week" in Chinese; and five months is "five GE month." Interestingly, in speaking "May" and "five months" in Chinese, the only difference between them is that the former has no unit GE (five month), whereas the latter has the unit GE (five GE month). It is not surprising that many Americans confuse May and five months when they learn to speak Chinese.

Many people suggest that mathematics is culturally neutral, that "numbers are numbers," and that mathematical operations should function the same across cultures, but not everyone agrees with this assertion (e.g., Antonouris, 1988). Because of the differences in languages with respect to number (i.e., between English and Chinese-based languages), one may argue that the variability in mathematical performance may also be due to differences in the representation of numbers among cultures. Thus, Asian children have an advantage based on how their language is linked with the number system that, in turn, may work to their advantage in later mathematics learning. In fact, Song and Ginsburg (1987) did show that Korean children learned addition and subtraction concepts more rapidly than American children in the early school years.

Curriculum. In viewing the importance of the curriculum in student learning and teacher training, one reasonable hypothesis is that the differences in mathematical performance are due, in part, to the differences in the curriculum in each country (Ginsburg & Clift, 1990; Jackson, 1992). Findings of curriculum differences from cross-national studies seem to support this hypothesis. For example, an obvious curriculum difference is that Asian countries, such as China and Japan, use a unified national curriculum in school, regardless of the students' academic progress. In the United States, there is no national curriculum in mathematics. This difference may affect the relative performance of students in these countries.

With respect to the content of the curriculum, Asian curricula seem more comprehensive than those of the U.S. For example, findings from the SIMS showed that both the Japanese intended curricula (the intentions of the school curriculum) and the implemented curricula (the content actually implemented) included more mathematical content than the curricula in the U.S. (Travers & Westbury, 1989). A comparative study conducted by Harnisch et al. (1985) revealed that particular mathematics courses taken also play a major role in explaining performance differences between Japanese and U.S. students.

In order to construct a mathematics achievement test for use in a cross-national study, Stigler, Lee, Lucker, and Stevenson (1982) analyzed elementary school textbook series used in Japan, Taiwan, and the U.S. to determine the grade level at which various concepts and skills were introduced. Across the elementary school textbook series from Japan, Taiwan, and the U.S., a total of 320 concepts and skills were identified. Of the 320 concepts and skills, 64% appeared in all three curricula, 78% appeared in the U.S. series, 80% in the Taiwanese, and 91% in the Japanese. The Japanese curriculum incorporated more concepts and skills than the curricula of Taiwan and the U.S. Although the U.S. and Taiwan series do not differ greatly in the percentage of total topics included, only 68% of the concepts and skills are common in the two series.

Grade level placement of topics is another area that may reveal national differences. Fuson, Stigler, and Bartsch (1988) investigated the grade placement levels at which addition and subtraction are introduced in the primary grades. They reported that addition and subtraction are introduced relatively late in U.S. mathematics textbooks as compared to textbooks in the former Soviet Union, Taiwan, China, and Japan. However, there is a remarkable uniformity in the placement of addition in the curricula of Japan, China, the former Soviet Union, and Taiwan. Stigler, Lee, Lucker, and Stevenson's (1982) analysis of the textbook series used in Japan, Taiwan, and the U.S. also determined the relative rate at which new concepts and skills were introduced. Throughout the six years, only 26 of the 320 topics were introduced during the same semester in all three countries. The Japanese curriculum, in general, introduced concepts and skills earlier than the curricula of Taiwan and the U.S.

Comparisons between the Addison-Wesley *Mathematics* textbook series (grades 1–6) (Eicholz, O'Daffer, Charles, Young, & Barnett, 1987) and comparable Chinese elementary school mathematics textbooks (Division of Mathematics of People's Education Press, 1987), also revealed differences in grade level placement of topics and organization of content (Cai, 1991). Substantial differences existed in the grade level at which decimals and fractions were introduced in the Chinese and American curricula. The Chinese textbooks introduced decimals (in the third grade) before fractions (fourth grade), whereas the Addison-Wesley *Mathematics* series introduced fractions (first grade) before decimals (third grade). Resnick et al. (1989) studied how different textbook orderings of topics influenced the probability that certain classes of student errors on solving decimal tasks would appear among the performance of students from France, Israel, and the U.S. They found that

> the fact that French children by and large avoided the errors associated with the fraction rule and instead seem to pass directly to the use of the correct decimal comparison rule might seem to suggest a superiority of the French curriculum sequence (which is shared by several other countries) in which decimal fraction instruction precedes ordinary fraction instruction by a substantial period of time. (pp. 25)

A larger percentage of content is repeated in American curricula than in Chinese and Japanese curricula. Flanders (1987) examined the percentage of new content introduced at each grade level (K–8) in three U.S. mathematics textbook series. He reported that the average percentage of new content in the three series ranged from about 40% to 65% at each grade level and that much of the new content is introduced at the end of the year. Cai (1991) coded Chinese elementary school textbooks using the Flanders coding method and found that over 95% of the content is new in each grade level (grades 1–6) in Chinese textbook series. In contrast to the U.S., in Chinese textbooks the old content is primarily found in the review section at the end of each textbook. Figure 2 shows the percentages of new content introduced at each grade level in the Chinese national-unified elementary mathematics textbooks and the Addison-Wesley *Mathematics* series.

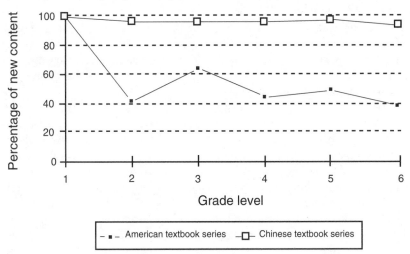

Figure 2. Percentages of new content in Chinese textbooks and Addison-Wesley Mathematics

Stevenson and Bartsch (1992) surveyed the content of American and Japanese elementary and secondary school textbooks and found that many concepts and topics were introduced in earlier grades in the Japanese textbooks than in the American textbooks. As in the findings of Flanders (1987) at grade levels K–8, a great deal of repetition of concepts was found in American secondary school textbooks. For example, more than 70% of the concepts in American secondary school textbooks were repeated at least once, nearly 25% were repeated twice, and 10% were even repeated three times. However, only 38% of the concepts in Japanese secondary school textbooks were repeated once and only 6% were repeated more than once. The fact that the U.S. curriculum contained so much old content may contribute to students' boredom if they sense that they do the same thing in mathematics year after year (Flanders, 1987).

The findings concerning curriculum differences are consistent with the exposure hypothesis: Students in Asian countries receive exposure to more mathematics than their counterparts in the United States. Westbury (1992) has provided some direct evidence in support of the exposure hypothesis by examining both curriculum and achievement in Japan and the United States in grade 7–8 algebra and grade 12 elementary functions and analysis. In an examination of the SIMS data, he found that the lower achievement of the U.S. students may have been a consequence of a curriculum not as well matched to the SIMS tests as the curriculum of Japan; further, when the American curriculum was congruent with the test items and with the curriculum of Japan, the achievement of American students was similar to that of the Japanese students.

Teachers. Teachers have an important effect on the ways in which students learn mathematics and on students' achievement in the subject (Grouws, Cooney, & Jones, 1988; Wittrock, 1992). One may argue that differences in student performance are due partially to differences among the teachers in each country. Teachers' ages, their amount of formal education, and their teaching experience do not seem to contribute to the performance differences in mathematics. However, differences between teachers' beliefs and teaching behaviors may account for performance differences in mathematics.

Significant differences have been found in teachers' beliefs about the best methods for teaching mathematics (Stigler & Perry, 1988). American teachers tend to believe that young children need concrete experiences in order to understand mathematics, at times asserting that concrete experiences will automatically lead to understanding. Chinese and Japanese teachers, however, apparently believe that even young children can understand abstraction and that concrete experience only serves to mediate an understanding of abstract mathematics. Chinese and Japanese teachers also believe that the more a student struggles, the more the student can learn; so teachers in China and Japan usually pose difficult problems to challenge students. U.S. teachers, in contrast, tend to pose problems "that will reinforce the idea that mathematics problems should be solvable in a single, insightful motion" (Stigler & Perry, 1988). Furthermore, U.S. teachers reported that mathematics was rather easy to teach, whereas the Japanese and the Chinese teachers stated that it was difficult to teach.

Teaching style also varies across cultures. It is not uncommon for Asian teachers to organize an entire lesson around the solution to a single problem. Asian teachers also often ask students to generate multiple solutions, while many American school students are rarely asked to do so (Stigler & Stevenson,

1991). More specifically, the one-problem-multiple-changes (OPMC) instructional approach has been regularly used in Chinese and Japanese classrooms (e.g., Cai, 1987; Hashimoto, 1987; Zhong, 1988). In OPMC, the lesson is organized around a single problem posed at the outset of the lesson, and the problem is changed multiple times along various dimensions as the lesson progresses. Henningsen and Cai (1993) suggested that a teacher can achieve coherent instruction by using the OPMC instructional approach.

Stigler et al. (1987) observed both students' and teachers' classroom behaviors in Japan, Taiwan, and the United States. They used five categories to code teaching behavior during each of the 10-second observational intervals: direct teaching of mathematics, giving directions, asking questions about mathematics to an individual student, asking questions about mathematics to a group (or to the class as a whole), and asking questions about topics unrelated to mathematics. Stigler et al. (1987) found that little time was spent in American classrooms imparting information; a moderate amount of time was spent in Japanese classrooms; much more time was spent in Chinese classrooms. Asian teachers are more likely to question the class as a whole, whereas American teachers are more likely to question individual children. Asian teachers also appear to spend less time asking nonacademic questions and giving directions than American teachers.

Students' attitudes. It has been consistently found that students' attitudes toward mathematics are related to their performance (Aiken, 1979; Cao & Cai, 1989; McLeod & Adams, 1989). A variety of students' attitudes have been surveyed and compared in cross-national studies, including attitudes toward school in general and attitudes toward mathematics in particular.

Stevenson and Lee (1990) indicated that Chinese and Japanese children regarded elementary school more positively than American children. American children did not especially like school; nevertheless, they believed that they were doing well in school, that they were meeting the expectations of their parents and teachers, and that they were bright students with high potential for academic work. Asian children rarely thought that they did as well in school as their parents and teachers expected. Chinese and Japanese children tended to believe that effort was the most important characteristic for academic success, whereas American children tended to believe that intelligence was most important for academic success. These beliefs might help clarify why American children were academically less successful than the Asian children. Stevenson and Lee (1990) argued that the very high self-perception and confidence of the American children might prevent them from acknowledging the need to work hard.

Findings from the SIMS showed that Japanese students thought the mathematics tested was more difficult than American students did, and they liked it less than students from the United States. Japanese students also thought that mathematics was more important than American students did. The IAEP (Lapointe et al., 1992) reported that most students from China and Korea did not view mathematics as a subject to be memorized, whereas more than half of the American students believed that learning mathematics is mostly memorizing.

Time spent on mathematics in and out of school. Results from research studies (e.g., Anderson & Postlethwaite, 1989; Frederick & Walberg, 1980) suggest that the more time spent on teaching and learning, both in and out of school, the greater the achievement of students will be. Thus, it seems also reasonable to hypothesize that differences in student performance are due, at least in part, to the differences in time spent on the teaching and learning of mathematics among nations. Data from some cross-national studies (e.g., Stevenson & Lee, 1990) provided evidence that supports that hypothesis. However, data from some other studies (e.g., McKnight et al., 1987) suggest that the relationship between mathematical performance and the time spent on mathematics is not simply linear. Previous cross-national studies surveyed (1) the amount of time spent on all subjects in school, (2) the amount of time spent on mathematics in school, (3) the amount of time spent on homework in all subjects, and (4) the amount of time spent on mathematics homework. The sections that follow summarize important findings from each of these topics.

With respect to the amount of time spent on all subjects in school, previous studies consistently showed that Asian students had a longer school year, whether measured in days or in hours, than U.S. students. For example, the data from the SIMS (McKnight et al., 1987; Robitaille & Garden, 1989) showed that in the secondary school the number of school days per year was 243 days for Japan but only 180 days for the U.S. The results from the second IAEP were similar, showing that the U.S had fewer school days per year (178) than China (251), Korea (222), and Taiwan (222). Stevenson and Lee (1990) reported that the hours spent in school by first-grade students were roughly comparable in Japan, the U.S., and Taiwan, ranging from 1 044 to 1 162 hours per year. By the fifth grade, however, the differences were more pronounced: fifth-grade children attended school for 1 044 hours per year in the U.S., 1 655 hours in Taiwan, and 1 466 hours in Japan.

Thus, compared to Korea, Japan, China, and Taiwan, the United States has the shortest school year. However, it cannot be automatically assumed from the length of the school year that U.S. schools devote the least amount of school time to mathematics. The findings from some cross-national studies on this issue suggest that U.S. students do spend the least amount of school time on mathematics. For example, Stevenson and Lee (1990) reported that Chinese and Japanese students in the first and fifth grades were estimated to spend over twice as much school time on mathematics as U.S. students in these grade levels. In particular, for the first graders, the amount of school time spent on mathematics was 2.7 hours per week for U.S. students, 5.8 hours per week for Japanese students, and 4 hours per week for Chinese students. For the fifth-graders, the amount of school time spent on mathematics was reported as 3.4 hours per week for U.S. students, 7.8 hours per week for Japanese students, and 11.7 hours per week for Chinese students. They also indicated that not only did the U.S. students spend fewer days in school each year and fewer hours in school each day, but they also spent a lower percentage of school time participating in academic activities.

However, some other studies suggest that U.S. students spent as much as or even more school time on mathematics than students in some Asian countries. For example, although the results from the second IAEP showed that 13-year- old students from Korea, China, and Taiwan had longer average days of instruction each year in school than their U.S. counterparts, only the students from China had longer average minutes of instruction on mathematics in school each week (307 minutes per week) than U.S. students (228 minutes per week). In fact, U.S. students had more instruction in mathematics in school than the students from Korea (179 minutes per week) and Taiwan (204 minutes per week). Similarly, McKnight et al. reported that for population A (7th and 8th graders) the number of hours allocated to mathematics instruction in Japan was 101 hours per year, which is much less than the 144 hours per year allocated to mathematics instruction in the U.S. Similar findings were obtained for population B (12th graders) with respect to the yearly hours of mathematics instruction.

In contrast, Garden and Livingstone (1989) reported the official school days per year, periods per day, minutes per period, and proportion of time for mathematics based on the SIMS. On the basis of Garden and Livingston's data, the number of hours spent on mathematics in Japanese and U.S. school systems each year was calculated by using the following formula: official school days per year × periods per day × minutes per period × proportion of time for mathematics ÷ 60. Results based on this calculation suggest that U.S. and Japanese population A (7th and 8th graders) had a similar number of hours on mathematics instruction each year, but Japanese population B students (233 hours per year) had more hours on mathematics instruction than their U.S. population B counterparts (141 hours per year).

Thus, the yearly hours of mathematics instruction in the U.S. and Japan calculated from the official school days per year, periods per day, minutes per period, and proportion of time for mathematics reported in Garden and Livingstone (1989) are different from those reported in McKnight et al. (1987). The discrepancy might be due to their use of different data sources. The data reported in McKnight et al. were obtained directly from a questionnaire administered to teachers in each country. Previous cross-national studies have used a variety of methods, such as teacher questionnaires, student questionnaires, parent questionnaires, and official standard counts to collect data for time spent in and out of school. One of the major concerns with using questionnaire data, however, involves internal consistency. Results from questionnaires completed by teachers, students, and parents are not consistent. In particular, the estimates of time spent in school reported by teachers, students, or parents may differ from the official standard counts. For example, according to Japanese educators (e.g., Kazuo, 1989; Shigematsu, personal communication, May 12, 1995), the standard number of school hours for Japanese seventh-grade students is 87.5 hours, which is different from the estimates reported by teachers and students in the SIMS.

With respect to the time spent on homework, previous studies showed that Asian students spent more time on homework in all subjects than U.S. students. For example, Stevenson and Lee (1990) asked mothers and teachers to estimate

the amount of time spent on homework. The mothers in Taiwan and Japan estimated that their first-grade children spent over four times as much time each
week doing homework as that estimated by the U.S. mothers. Differences in the
time spent on homework among Chinese, Japanese, and U.S. students were
almost as large at the first grade as they were at the fifth grade. Differences were
not as dramatic in teachers' estimates as in mothers' estimates, but teachers also
estimated that children in Taiwan and Japan spent more time doing homework
than U.S. children. Moreover, when teachers were asked to rate how strongly
they believed homework to be of value to children, Stevenson and Lee (1990)
found that Chinese and Japanese teachers were more convinced that homework
was of value than U.S. teachers. Data from the second IAEP showed that a larger percentage of 13-year-old students from Korea, China, and Taiwan spent 2
hours or more on homework each day than U.S. students. However, data from
the SIMS showed that Japanese and U.S. students spent similar amounts of time
on homework in all subjects (Robitaille & Garden, 1989).

As far as the time spent particularly on mathematics homework is concerned,
data from the second IAEP showed that students from Korea, China, and Taiwan
spent more hours on mathematics homework each week than U.S. students. In
particular, 72% of the students from China reported that they spent 2 hours or
more on mathematics homework each week, but only 37% of the U.S. students
reported doing so. However, the data from the SIMS showed that Japanese students did not spend more time on mathematics homework than U.S. students. In
fact, population A students from the U.S. and Japan reported spending 3 hours
per week on mathematics homework; U.S. population B spent 4 hours per week
on mathematics homework, and Japanese population B spent 2 hours per week
on mathematics homework. It should be noted that the number of hours per
week spent on mathematics homework reported from SIMS is lower than what
was documented elsewhere (e.g., Becker, Silver, Kantowski, Travers, & Wilson,
1990; McKnight et al., 1987). For example, McKnight et al. (1987) indicated
that "many Japanese students spend a considerable amount of time studying
mathematics outside of regular school hours" (p. 52).

In summary, Asian students not only had a longer school year, but also spent
more time on homework in all subjects. Interestingly, with respect to the time spent
on mathematics instruction in school and on mathematics homework, the patterns
were somewhat different. The results from the previous cross-national studies
showed that U.S. students were not always spending less time in mathematics
instruction in school and mathematics homework than Asian students.

It is worth noting that previous cross-national studies focused mainly on the
allocated time rather than on "productive time." According to Walberg (1988),
productive time "is the time spent on suitable lessons adapted to the learner—in
contrast to 'engaged' or 'allocated' time, which may be futile if the content or
method of instruction is inappropriate for individual students" (p. 81). Although
the examination of allocated time in cross-national studies is a reasonable indicator of students' learning time, the time spent on suitable lessons adapted to the

learner may be a better indicator of students' learning. Moreover, Walberg (1988) argued that "the amount of time devoted to schooling in the U.S. has increased substantially during the 20th century, but it is difficult to argue that the time allocated is yet sufficient" (p. 84). Thus, although there was a lack of strong evidence to support the argument that U.S. students' poor mathematics performance was due to insufficient time, it is plausible that students may learn more by increasing their "productive time."

Home background variables. A strong relationship between home background variables and student performance is well documented (Robitaille & Garden, 1989). Recent research in cognition showed that students' experiences out of school have a substantial effect on their problem-solving skills (e.g., Resnick, 1987). Therefore, students' home environments should affect their academic achievement. Previous cross-national studies have surveyed the educational and occupational status of the parents, parental help, the study environment at home, parents' expectations of their children, and parents' beliefs about their children's future happiness and the way to achieve success. The educational and occupational status of the parents and the study environment at home were not found to contribute to performance differences in mathematics (e.g., Robitaille & Garden, 1989; Stevenson & Lee, 1990). However, parental help, parents' expectations of their children, and parents' beliefs about their children's future happiness and the way to achieve success do seem to be related to children's school work (e.g., Lapointe et al., 1992; Robitaille & Garden, 1989; Stevenson & Lee, 1990).

American parents were more likely to help their children with homework than to ask them about their mathematics classes. In contrast, Chinese parents were more likely to ask their children about their mathematics classes than to help them with their homework (Lapointe et al., 1992). Both Japanese and Chinese mothers appeared to believe that the route to future happiness is through hard work and high academic success, whereas American mothers gave greater emphasis to innate ability. American mothers appeared to be less interested in their children's specific academic achievement than in their children's general cognitive development, so they attempted to provide experiences that fostered cognitive growth rather than academic excellence. Although both Chinese and Japanese mothers emphasized diligence as a way to achieve success, American mothers stressed independence, innate ability, and acceptance of diversity. Chinese and Japanese mothers also held higher standards for their children's achievement than American mothers, and they gave more realistic evaluations of their children's academic, cognitive, and personality characteristics.

Conditions and requirements in schools are obviously different from those at home. Some students are more ready than others to deal with these conditions and requirements. Students' readiness or adaptive dispositions come from attributes they bring to school. According to Hess and Azuma (1991), students' adaptive dispositions include

(a) willingness to master skills and engage in tasks, whether appealing or not; (b) readiness to accept the curriculum of the school; (c) willingness to accept the rules of the school and authority of the teacher; (d) ability to concentrate and to persist in tasks and complete them on time; (e) readiness to monitor one's own behavior and performance, giving attention to detail and to quality of work; (f) ability to work independently outside of school or in the classroom while the teacher is engaged with other students; and (g) willingness to accept rules of social behaviors necessary for learning in groups. (p. 2)

Hess and Azuma (1991) have emphasized the importance of developing an adaptive disposition for students to be successful in school. They document extensively how Asian home and cultural support facilitates their children's development of adaptive dispositions.

Summary of contributing factors. Cross-national studies indicate that there are differences between Asian and American school children in mathematical performance and that differences in schooling, parental support, and student attitude contributed to Asian children's superior mathematics achievement. In particular, the most salient contributors to achievement differences are curriculum, teacher beliefs about mathematics teaching, teaching approaches, time spent on mathematics, student attitudes, parental expectations of children's achievement, and parental beliefs about the roles of effort and ability. The linguistic difference of numbers between Chinese-based language and the English language might be another factor that contributes to Asian students' superior mathematics achievement.

The identified factors discussed above should not simply be interpreted as causal factors that explain the differences in performance. At times, certain patterns of differences show no correlation at all with mathematical performance. For example, it is well known that Asian class size (over 40 students in each class) is larger than American class size (up to 30 per class). Still, class size may not be a factor that explains performance differences between American and Asian students. In other words, it is inappropriate to conclude that Asian students' superior performance in mathematics is necessarily related to large class size.

In the study reported here, no background information was collected to interpret the possible performance differences and similarities between American and Chinese students, because it was not the focus of the study. The contributing factors reviewed in this section, however, are helpful in understanding performance similarities and differences that may be observed between American and Chinese students.

Methodological Issues in Cross-National Studies

A cross-national study involves much more than collecting data in two countries and comparing the results (Farrell, 1979; Miller-Jones, 1989; Postlethwaite, 1987; Raivola, 1985). Sometimes, the differences in test scores may not reflect differences in the traits purported to be measured by the tests (Berry, Poortinga, Segall, & Dasen, 1992). There are two main methodological issues in a cross-national study: sampling methodology and data collection procedures.

Sampling methodology. According to Bracey (1992, 1993) and Rotberg (1990), one of the major problems with international comparisons is related to sampling. Both Bracey (1992, 1993) and Rotberg (1990) asserted that samples of students in some cross-national studies in mathematics were not representative of the respective populations. For example, in a comparative study conducted by Stevenson et al. (1990), the achievements of a sample of first- and fifth-grade students from Chicago and Beijing were compared on different mathematical topic areas. Bracey (1992) indicated that the student samples in Chicago and Beijing were "in no way" similar.

Rotberg (1990) selected from the SIMS data the two countries with the highest percentage of the age group taking mathematics in the twelfth grade and compared their data with that of the two countries with the lowest percentage of the age group taking mathematics in the twelfth grade. She concluded that "countries with a high proportion of young people taking twelfth-grade mathematics would rank relatively lower in the 12th-grade comparisons than they did in the eighth-grade comparisons, while countries that retain only a small, highly selected group in mathematics would rank relatively higher in the twelfth-grade comparisons" (p. 297). Therefore, the rank order of participating countries in SIMS is related to student selectivity, and an over-generalization of those findings is misleading.

Rotberg (1990) also indicated that in international comparisons each country's sample should represent the entire national distribution of the age group examined. Some studies, however, failed to select representative samples. For example, the international comparisons in the IAEP are "seriously biased because only the most prosperous regions or the most elite schools and students will be sampled in some of the participating countries" (Rotberg, 1990; p. 298). After they carefully reviewed five international achievement surveys of mathematics and science[3], Medrich and Griffith (1992) concluded that "it is not clear that comparable populations have been tested across participating countries. …[Therefore], key results of these studies … should be viewed cautiously because they are more likely indicative of achievement-related trends and patterns, rather than definitive and conclusive" (pp. 21–22).

Achieving sample comparability and representativeness is an important, but often elusive, goal. Obviously, sampling problems in cross-national studies are not easy to overcome. Two strategies have been commonly used to reach comparability in sampling: age-level and grade-level sampling. Both age-level and grade-level sampling have advantages and disadvantages (Medrich & Griffith, 1992), but grade sampling provides the opportunity to relate classroom characteristics to student performance in ways that would not be possible with an age-based sample.

Bradburn and Gilford (1990) provided criteria of the sample designs for the large-scale studies: "Each sample should be designed so as to support reasonably accurate inferences about an age or grade cohort, and the proportion of each

---

[3]The five surveys are the First International Mathematics Study (FIMS), the First International Science Study (FISS), the Second International Mathematics Study (SIMS), the Second International Science Study (SISS), and the First International Assessment of Educational Progress (IAEP-I).

cohort covered should be carefully estimated and reported. The sample should be designed to ensure it captures the range of individual, school, or classroom variation that exists in the nation sampled" (p. 25). For relatively small and localized samples, however, an in-depth analysis of students' thinking is needed. According to Bradburn and Gilford (1990), the information from these in-depth studies can also play an important role in educational research and policy development.

Data collection across cultures. Besides sampling concerns in international comparisons, researchers also voice concern over the data collection procedures that are often used. Interpretations of cross-national data must be carefully made to guard against the effects of cultural bias in the data collection. The first concern in the collection of data across cultures is the selection of the test items. Every attempt must be made to ensure the fairness of the selected items for the two nations' samples. In cross-national studies in mathematics, mathematical performance has generally been assessed on the basis of a set of tasks. It is assumed that mathematical performance can be measured from this set of tasks and that the levels of mathematical performance can also be represented by a numerical score. Different conclusions about mathematical performance might be reached from tests emphasizing different formats.

For example, consider two mathematics problem-solving studies (Baranes, Perry, & Stigler, 1989; Carraher, Carraher, & Schliemann, 1987) that used symbolic computational problems, textbook-like word problems, and situation problems. In the first study, Carraher et al. (1987) found that Brazilian children were more successful in solving textbook-like word problems and situation problems than in solving symbolic computational problems. They also found that Brazilian children solved the problems that were embedded in real-world situations more easily than those embedded in school-like situations, although they required the same computations. In the second study, Baranes et al. (1989) replicated the Carraher et al. study using U.S. students as subjects. Interestingly, they did not find a significant effect for context (symbolic computation problems vs. textbook-like word problems vs. situation problems). The difference between U.S. and Brazilian students on the symbolic computation tasks is larger than the difference on the textbook-like word problems and situation problems. The performance differences between U.S. and Brazilian students may be derived differently, depending on the tasks the comparison is based on. Thus, the nature of the tasks is a crucially important factor to consider in measuring students' mathematical performance.

Bracey (1992) questioned whether multiple-choice test items are the most appropriate format to measure students' mathematical performance. This concern is consistent with issues being raised in the current testing reform movement in the United States. One of the major criticisms of the multiple-choice test format is that it does not allow students to produce their own answers; to display the process used to obtain an answer; to explain the thinking or reasoning associated with their responses; or to exhibit alternative approaches to, or interpretations of, a problematic situation (Cai et al., in press-a; Magone et al., 1994;

Silver, 1992). Moreover, there is the difficulty of inferring underlying cognitive processes from performance on multiple-choice format testing (Cai et al., in press-a; Miller-Jones, 1989). An ideal cross-national study would involve an assessment of students' mathematical performance across a range of different, complementary test formats.

The second concern of cross-national data collection has to do with test administration. In order to have valid and reliable data that allow for varied comparisons across cultures, an adequate and consistent test administration procedure is necessary. A critical issue in test administration is *translation equivalence*. The use of a test in a language group other than the one for which it was originally designed has often led to a concern about translation equivalence. A careful analysis of the effect and quality of translation is needed, both for test items and for administrative directions. Two methods have been suggested to ensure translation equivalence (Berry et al., 1992; Brislin, 1986). The first way to attain translation equivalence is to translate from the original language to the target language, follow this with independent back-translation, and then examine the precision of comparability between the back-translation and the original wording (Berry et al., 1992). In many cases, however, it is almost impossible to translate a word from one language to a completely equivalent word in another language. Brislin (1986) suggested a second way to attain translation equivalence: Before the test items were formally used, they should be piloted to check whether students in both languages interpret the test items and directions in the same way.

Bracey (1992) also expressed concern over the comparability of students' attitudes toward testing in varied countries. American students, for example, might not take testing as seriously as Asian students (Bracey, 1992). The findings of a comparative study of task persistence by Blinco (1991) appear to support Bracey's concern. In the Blinco (1991) study, Japanese and U.S. children were given the first stage of a manipulative puzzle-like game that had a series of increasing stages of difficulty. The final goal was for the child to master the task of correctly assembling each stage of the game. The findings were that Japanese elementary school students persisted significantly longer than American students. Such differences are of concern to those trying to establish equivalent testing conditions in order to ensure the comparability of performance data.

CONTRIBUTIONS OF COGNITIVE PSYCHOLOGY TO CROSS-NATIONAL STUDIES IN MATHEMATICS

Cognitive Psychology and Measurement of Performance

Cognitive psychology is the study of how people mentally represent and process information (Anderson, 1985; Simon, 1979, 1989). Cognitive processes such as attention, perception, memory, learning, reasoning, and problem solving are often the focus of inquiry for cognitive psychology. Generally, cognitive psychologists view the human mind as an information-processing system.

Therefore, the cognitive approach is based on a human-computer analogy. Just as a computer system contains hardware and software, so the human mind is viewed as having hardware (i.e., fundamental memory capacities and processing characteristics) and software (i.e., knowledge and its organization). Cognitive psychology studies the structure, function, and interactions between these aspects of the mind.

Cognitive psychologists have pursued different approaches to understanding human mental processes and knowledge structure. Sternberg (1984, 1991) classified cognitive approaches into four different categories: cognitive correlates, cognitive components, cognitive training, and cognitive contents. Cognitive approaches have the potential to improve the measurement of performance in general and the measurement of mathematical performance in cross-national studies in particular. In fact, cognitive approaches have already been applied in educational measurement (Bejar, Embretson, & Mayer, 1987; Snow & Lohman, 1989).

The cognitive-correlates approach is based on an analysis of human mental abilities to perform tasks that contemporary cognitive psychologists believe measure basic human information-processing abilities. The cognitive-components approach is based on an analysis of human mental abilities to perform tasks believed to measure higher-level human information-processing abilities. Bejar et al. (1987) used the computer analogy to distinguish the cognitive-correlates from the cognitive-components approaches. They described the former as the analysis of the hardware of the information-processing system, the latter as the analysis of the software of the information-processing system. Sternberg (1991) described the differences between the cognitive-correlates and the cognitive-components approaches while admitting that the distinction between them was not entirely clear-cut:

> If one were willing to accept a continuum of levels of information processing extending from perception, to learning and memory, to reasoning and complex problem solving, cognitive-correlates researchers would tend to study tasks measuring skills at the lower end of the continuum, whereas cognitive-components researchers would tend to study tasks measuring skills at the higher end of the continuum." (p. 372)

Educational measurement of performance in cross-national comparative studies focuses primarily on an analysis of the software of the information-processing system with respect to mathematical problem solving. The cognitive-components approach involves not only the higher level of skills of performing a task, but also the required knowledge, processes, and strategies of performing the task. Therefore, the cognitive-components approach seems to be useful for the kind of task analysis needed in a cross-national study. The cognitive-training approach seeks to infer identities of cognitive processes through an analysis of the effects of training. This approach may not be used directly in a cross-national study, but it may be used indirectly to provide suggestions and recommendations based on the outcomes of the comparison.

The cognitive-contents approach seeks to compare the performances of experts and novices in complex tasks such as physics and mathematics problems. Sternberg (1991) suggested that the cognitive-contents approach might supplement psychometric tests with complex learning or problem-solving tasks that elicit an examinee's knowledge base, the way in which knowledge is mentally represented, the character of cognitive strategies, and the way in which cognitive strategies are employed. The cognitive-components approach can be used to conduct the analysis of student responses in a cross-national study, in which students' knowledge and cognitive strategies are examined.

In fact, researchers have proposed the applications of cognitive theories and research methods to educational assessments (e.g., Frederiksen, Glaser, Lesgold, & Shafto, 1990; Glaser, 1987; Pellegrino, 1992; Ronning, Glover, Conoley, & Witt, 1985; Snow & Lohman, 1989). In particular, cognitive psychologists have developed tools for analyzing the knowledge and cognitive processes required for performance in complex tasks. Some of these tools have been used to assess educational achievement.

Mayer (1987) developed a model for analyzing cognitive components in solving word problems. His model was based on the assumption that the two major phases of problem solving are (1) representing the problem and (2) searching for a means to solve the problem. In order to represent a problem, a student must be able to translate each sentence of the word problem into an internal representation such as an equation and be able to put the elements of the problem together into a coherent whole. Cognitive research has suggested that the breakdown in linguistic comprehension and the lack of schemas for problem types are the source of many difficulties in problem solving (Mayer, 1987). In order to search for a means to solve a problem, the student must also be able to plan and find an adequate algorithm and then flawlessly execute the algorithm. In Mayer's model, four cognitive components involved in solving mathematical word problems were classified and analyzed: translation, integration, planning, and execution. Translation and integration involve the representing phase of problem solving, and planning and execution involve the searching phase. This model has been used to assess students' general cognitive aspects of mathematical problem solving on the Scholastic Aptitude Test (SAT) (Bejar et al., 1987) and in a cross-national study (Mayer et al., 1991). These uses demonstrated the effectiveness of the model in measuring students' mathematical performance.

Advances in cognitive psychology have also contributed to identifying dimensions of measuring student performance (Glaser, Lesgold, and Lajoie, 1985; Royer et al., 1993) and the ways student achievement is measured (Lesgold, Lajoie, Logon, & Eggan, 1990). For example, Glaser et al. (1985) identified eight dimensions for a cognitive approach to the measurement of achievement: (1) knowledge organization and structure, (2) depth of problem representation, (3) quality of mental models, (4) efficiency of procedures, (5) automaticity to reduce attention demands, (6) proceduralized knowledge, (7) procedures for theory change, and (8) metacognitive skills of learning. These should be fruitfully applied to a cognitive approach to the measurement of performance.

Cognitive research has revealed that "learning is not simply a matter of the accretion of subject-matter concepts and procedures; it consists rather of organizing and restructuring of this information to enable skillful procedures and processes of problem representation and solution" (Glaser et al., 1985, p. 41). According to Glaser et al. (1985), there are different levels of knowledge organization and different degrees of cognitive processes and procedural skills involved in different stages of learning. The different degrees of performance highlight the different stages in the course of learning.

The performance measurement defined by Glaser et al. (1985) needs to be shaped by theories of the acquisition of subject-matter knowledge. In particular, Glaser and his associates "anticipate that theories of subject-matter acquisition will suggest both general indicators of competent performance, and also specific indicators dependent upon the nature of the knowledge and skill being assessed" (1985, p. 81). The eight dimensions of measuring performance listed above are an appropriate reference for developing a framework for doing cognitive-based cross-national studies in mathematics, because along these dimensions the competent performance of mathematical problem solving can be indicated.

Conceptual Framework for Measuring Mathematical Performance

Recently, the Mathematics Sciences Education Board (MSEB, 1993) and National Council of Teachers of Mathematics (NCTM, 1995) released conceptual guides for mathematics assessment. These documents, like the *Curriculum and Evaluation Standards for School Mathematics* (NCTM, 1989) and the *Professional Standards for Teaching Mathematics* (NCTM, 1991), indicate that "Mathematics is no longer just a prerequisite subject for prospective scientists and engineers but is a fundamental aspect of literacy for the twenty-first century" (MSEB, 1993, p. 2). Contemporary views of mathematics education emphasize the importance of thinking, reasoning, conceptual understanding, problem solving, and communication. As the views of mathematics and mathematics education have changed, the measurement of mathematical performance has to be changed to ensure alignment with these new visions of mathematics and mathematics education (Romberg, Zarinnia, & Collins, 1990; Romberg, Wilson, Khaketla, & Chavarria, 1992). In a cognitive-based cross-national study, the framework for measuring mathematical performance should be built on a new view of mathematical performance and advances in cognitive psychology and performance assessment in mathematics (Broadfoot, Murphy, & Torrance, 1990; Eckstein & Noah, 1991; Gifford & O'Connor, 1992; Niss, 1993).

The traditional ways of measuring mathematical performance generally fail to directly measure higher-level skills such as thinking, reasoning, conceptual understanding, problem solving, and communication (Putnam, Lampert, & Peterson, 1989; Resnick & Resnick, 1992; Silver, 1992). For example, Romberg et al. (1991) examined six widely used standardized tests in the U.S. to determine whether or not they are appropriate instruments for assessing the content, process, and levels of thinking called for by the NCTM *Standards* (1989). Their findings showed that the tests place too much emphasis on computational procedures and routine word problem solving and not enough emphasis on conceptual understanding and nonroutine problem solving. Obviously, it is not sufficient to measure mathematical performance on the basis of computation and routine verbal problem solving, although these are important skills in mathematics. Thus, the measurement of mathematical performance should include higher-level skills, such as reasoning, communication, and problem solving.

In the past, test scoring has basically provided quantitative information regarding student performance (Sternberg, 1991). Rarely have test scores reflected the qualitative or cognitive aspects of information processing. However, increasing emphasis on a cognitive analysis of performance has shifted attention more to the cognitive or qualitative aspects of information processing: strategies for problem solving, modes of representation, and the like. In particular, Sternberg (1991) proposed that "test scores of the future should reflect this qualitative emphasis, as well as providing quantitative data of the kind used in the past" (p. 388). The scoring of mathematical performance should align with Sternberg's proposal for future testing.

In particular, a detailed cognitive analysis of students' work in solving mathematical problems is needed in order to examine cross-national similarities and differences. The cognitive analysis of students' work in solving mathematical problems should be conducted on the basis of critical cognitive aspects of the process including solution strategies, mathematical misconceptions, mathematical justifications, and modes of representation. Indeed, these cognitive aspects have been identified as important and significant dimensions in cognitive psychology in general (e.g., Royer et al., 1993) and in mathematical problem solving in particular (e.g., Charles & Silver, 1989; Silver, 1987).

Solution strategy. A strategy is a plan or a way of attack for achieving a goal (Simon, 1979). Cognitive psychologists have distinguished two types of cognitive strategies (Anderson, 1987): general cognitive strategies (or weak strategies) and domain-specific cognitive strategies (or strong strategies). General strategies for problem solving such as brainstorming, means-end analysis, reasoning through analogy, the use of worked examples, working backwards, and working forwards can be applied to problems in many different domains. The domain-specific strategies such as looking for a pattern can only be applied to problems in a particular domain such as mathematics.

Cognitive psychologists have distinguished between experts and novices with respect to the general cognitive strategies and domain-specific cognitive strategies they use in problem solving. Psychologists have found that experts and novices employ different strategies in solving problems, with experts appearing to use more adequate strategies than novices. For developmental psychologists (e.g., Bjorklund, 1990), children's strategies in solving problems are closely related to their cognitive development. Psychologists have also found that older children use more sophisticated strategies than younger children and that cognitive strategies are mediators of cognitive development.

In mathematics, from simple counting tasks to complex mathematical problem solving, individuals use different strategies (Carpenter, Moser, & Romberg,

1982; Schoenfeld, 1979, 1985; Siegler & Shrager, 1984). Individual differences in solving mathematical problems can sometimes be understood in terms of differences in the use of different strategies. According to Glaser (1987), proficiency in solving mathematical problems is dependent on the acquisition, selection, and application of both content-specific strategies and general problem-solving strategies. There has been a longstanding tradition of teaching problem-solving strategies in the mathematics education community (Polya, 1954, NCTM, 1989). A number of researchers (e.g., Pressley & Associates, 1990; Schoenfeld, 1979) found that cognitive strategy instruction greatly improved students' performance on many school subjects, including mathematics. Thus, competence in using appropriate problem-solving strategies reflects students' degrees of performance proficiency in mathematics. This implies that the examination of the strategies that U.S. and Chinese students apply and the success of those applications can provide information regarding their mathematical thinking and reasoning.

Mathematical misconceptions. The study of misconceptions is also related to the study of human cognition, particularly to the study of the nature of competent performance and learning. Cognitive psychologists (e.g., Glaser, 1987) have concluded that conceptual changes are an important part of learning—that is, the changes that occur as knowledge and complex cognitive strategies are acquired. Children, and even adult learners, start with a "naive theory" (Kuhn, 1989). When appropriate instruction has taken place, students may modify their naive theories on the basis of new information and, as a result, form "new theories." Nevertheless, after instruction, these naive theories may persist. The study of misconceptions provides information with respect to the characteristics and treatment of misconceptions and the extent to which students modify their misconceptions.

In mathematics, many studies have been conducted to detect students' errors by examining their understanding of counting principles in numerical reasoning (Greeno, Riley, & Gelman, 1984); their understanding of principles that underlie place value concepts (Resnick, 1982; Resnick et al., 1989); their acquisition of arithmetic facts and procedures (Leinhard et al., 1992; Brown & VanLehn, 1980; Resnick, 1984; VanLehn, 1983); their knowledge of, and tactics for, solving arithmetic word problems (Kintsch & Greeno, 1985; Riley, Greeno, & Heller, 1983); their ability to engage in situation-based reasoning (Silver & Shapiro, 1992); and their understanding of important mathematical concepts in particular content areas such as algebra (Sleeman, Kelley, Martinak, Ward, & Moore, 1989; Steinberg, Sleeman, & Ktorza, 1991), statistics (Allwood, 1984), or advanced mathematics (Tall, 1991). The research on the diagnosis and categorization of student error patterns in mathematics provides a basis for informed diagnosis of student understanding and misunderstanding in mathematics learning. As Glaser (1987) stated,

> It will become easier to identify the incomplete knowledge and procedures and incomplete conceptual understanding that contribute to weak performance and can be remedied in the course of instruction. We are able to appraise the knowledge that

reveals degrees of competence and that determines functional differences between superficial and more lasting achievement. (p. 332)

These previous investigations demonstrated the value of error analysis in capturing students' understanding of mathematical knowledge. The examination of students' misconceptions provides an indication of their proficiencies in mathematical problem solving and reasoning.

Mathematical justification. NCTM (1989) suggests that the study of mathematics should emphasize reasoning where "students can justify their answers and solution processes" (p.29), "make and evaluate mathematical conjectures and arguments," and "validate their own thinking" (p. 81). Studies by Senk (1985, 1989) showed not only that students have great difficulty in providing sound mathematical justifications when solving geometric proof problems but also that the degree of difficulty was related to students' levels of geometric thought.

Mathematical justification is related to communication, which is one of the emphases in NCTM's standards documents (NCTM, 1989, 1991). Since mathematics is viewed as a fundamental aspect of literacy, communication should be a central part of mathematics education. With this increasing awareness of the importance of communication in both instruction and assessment (Cai et al., in press-a; MSEB, 1993; NCTM, 1991, 1995; Silver & Cai, 1993), it is imperative that mathematical communication be an important dimension in assessing students' mathematics proficiency. For example, MSEB (1993) proposed the requirement of communication in task development and the evaluation of students' responses.

In solving open-ended mathematical problems, students are asked to justify their solutions. The examination of students' mathematical justifications provides important information about their communication skills. Student mathematical justifications are judged in terms of their soundness. Soundness refers to (1) whether the justification providing support is acceptable or correct and (2) whether the justification providing support is complete (Voss, Perkins, & Segal, 1991).

Modes of representation. Newell and Simon (1972) treated a problem solver as an information-processing system that creates problem representations and searches selectively through intermediate situations, seeking the goal situation and using heuristics to guide the search. More specifically, in solving a problem, the problem solver operates within a problem space. A problem space is a set of points or nodes, each of which represents a particular knowledge state. A knowledge state is the set of things the problem solver knows or postulates when he is at a particular stage in his search for a solution. Problem-solving activity can be described as a search through the space of knowledge states, until a state is reached that provides the solution to the problem. In solving a problem, a solver first needs to establish a representation of the problem. The problem representation includes the initial state of the problem (the "givens") and the goal state of the problem. The development of the problem representation largely influences the problem solutions. Chi, Feltovich, & Glaser (1981) found that experts tend to represent problems on the basis of concepts and principles, whereas novices tend to recognize and attend to the surface features of the problems.

The problem representation that cognitive psychologists refer to is also called a mental model or an internal representation. Goldin (1987, 1992) proposed five kinds of internal cognitive representational systems that students might employ during mathematical problem solving: (1) a verbal/syntactic system, (2) a visual-spatial system, (3) a formal notational system, (4) a system for heuristic planning and executive control, and (5) an affective system. In many cognitive studies of problem representation, the problem solvers' internal representations were inferred through their verbalization and external work. Modes of representation are the external representations of students' solution processes, which reflect their mathematical thinking. The examination of the modes of representation reveals the ways in which students process a problem and reflects the ways in which students communicate their mathematical ideas and thinking processes.

3. METHOD

SAMPLES

The subjects for this study were 425 sixth-grade Chinese students and 250 sixth-grade U.S. students. Boys and girls were about evenly distributed in the Chinese sample (224 girls and 201 boys) and in the U.S. sample (119 girls and 131 boys). The investigator had three potential locations in China as research sites: Beijing, Lanzhou (Gansu province), and Guiyang (Guizhou province). Beijing, the capital city of China, has many characteristics that make it unique compared to other cities in China. Therefore, Lanzhou and Guizhou were chosen as research sites, since they are more representative of China as a whole. In particular, the Chinese sample was chosen to include students from common schools (240) and from key schools (185) in Lanzhou and Guizhou cities. In general, Chinese key schools are highly selective schools that offer a high-quality educational program. However, no official labels exist in China to distinguish key schools from common schools. In this study, common or key schools were informally identified by a group of Chinese school principals and mathematics education professors. The inclusion of students from common and key schools ensured a somewhat more representative sample of Chinese students. It should be indicated that the sample from Lanzhou and Guizhou are from urban settings.

After the Chinese sample was chosen, the U.S. sample was selected. The U.S. sample was drawn from the Pittsburgh (Pennsylvania) metropolitan area. It was convenient to choose the U.S. sample from Pittsburgh, because the investigator was living there. The demographic characteristics of Pittsburgh also made it desirable, and the urban environment in Pittsburgh is similar to that in Lanzhou and Guizhou. Finally, Pittsburgh was also chosen because it offered a readily available sample of private-school students.

The choice to use students from private schools was viewed as desirable because the social and institutional backgrounds of students in U.S. private schools are more similar to those of Chinese students than those of students in U.S. public schools. For example, U.S. students who attend private schools are more likely than those who attend public schools to live in traditional and stable families (Cookson, 1989; James & Levin, 1988), because the frequency of divorce and single-parent homes is lower among private-school parents than among public-school parents (Cookson, 1989). In addition, private-school parents tend to have higher educational expectations of their children than public-school parents. Likewise, students who attend private schools have higher educational aspirations than students in public schools (Cloeman & Hoffer, 1987). Cloeman and Hoffer (1987) found that 76% of Catholic and 70% of other private-school ninth graders planned to attend college, whereas less than 60% of public-school ninth graders planned to attend college. These characteristics of U.S. private-school students are quite similar to those of Chinese students. The vast majority of Chinese students are from two-parent families. Chinese parents have very high expectations of their children. For example, a survey of junior high school Chinese students showed that over 75% of them said that their parents expected them to go to college (Cao & Cai, 1989).

There are other features of private schooling in the U.S. that made it likely that private-school students would constitute a good comparison group for Chinese students. In the U.S., private-school teachers tend to be very dedicated to their schools and children (Cookson, 1989). Teachers in China are also very dedicated (Tian & others, 1989). Private-school teachers in the United States are less likely than public-school teachers to have an advanced degree (Cookson, 1989). The survey of the teachers for the third-level students in junior high schools (ninth graders) showed that only 37% of Chinese mathematics teachers had college-level education (Tian & others, 1989). Although U.S. teachers in private schools have more formal education than Chinese teachers, the gap of formal education between Chinese teachers and U.S. private-school teachers is smaller than that between Chinese teachers and U.S. public-school teachers.

Programs in U.S. private schools are generally not vocational in nature but place an emphasis on academics, which is very similar to programs in Chinese regular schools. In the United States, private-school students take more hours in mathematics and spend more time on their homework than students in public schools (Cookson, 1989). Several comparative studies (e.g., Chen & Stevenson, 1989; Lapointe et al., 1992) have shown that Chinese students generally have more hours in mathematics each week and spend more time on their homework than U.S. students. Therefore, the time spent on mathematics in and out of school also suggests that U.S. private schools are more like Chinese schools than are U.S. public schools.

Eight private schools in Pittsburgh volunteered to participate in the study. Among these eight schools, four are exclusive private schools that have high academic reputations in the Pittsburgh area, and the other four are nonexclusive in status. In the sample, then, the exclusive U.S. private school students correspond roughly to the Chinese key-school students, and the nonexclusive U.S. private school students correspond roughly to the Chinese common-school students. Having U.S. students from both exclusive and nonexclusive private schools and Chinese students from both key schools and common schools is an attempt to have roughly equivalent samples. In total, 117 U.S. students from the four exclusive private schools and 93 from the four nonexclusive private schools were chosen. Since the ratio of U.S. exclusive private-school students to nonexclusive private-school students (117 to 93) was not close to the ratio of Chinese key-school students to Chinese common-school students (185 to 240) (i.e., the proportion of exclusive private-school students in the U.S. sample is larger than that of key-school students in the Chinese sample), an additional 40 U.S. students from a suburban public school in the Pittsburgh area were added to the U.S. sample. On the basis of the best available information from informal sources that are familiar with public schools in the Pittsburgh area, it was judged that the educational value and program in this public school was comparable to that in the nonexclusive private schools. Thus, the addition of these students to the U.S. sample helped to ensure that there would be a similar proportion of exclusive private-school students in the U.S. sample and key-school students in the Chinese sample.

There are some social and institutional background factors that need to be considered. For example, U.S. schools, especially private schools, generally have fewer students than Chinese schools. Some of the U.S. private schools are Christian and Catholic schools. The samples in this study are not nationally representative samples, but they are roughly equivalent samples. Only relatively small and localized samples were selected for this cross-national study because of the limited time and resources available, but the depth of the cognitive analysis of students' mathematical problem solving offsets this limitation. As Bradburn and Gilford (1990) suggested, for small in-depth cross-national studies, relatively small and localized samples in a small number of sites are acceptable. The information from these in-depth studies can also play an important role in educational research and policy development.

Including exclusive and nonexclusive school students in the U.S. sample and key- and common-school students in the Chinese sample was done to ensure roughly equivalent national samples, though neither sample is representative of either country, and students were not randomly chosen. However, the comparisons between exclusive and nonexclusive school students in the U.S. sample, between key- and common-school students in the Chinese sample, and between exclusive-school students in the U.S. sample and key-school students in the Chinese sample are not the focus of this study. All data analyses in this study were based on the total U.S. and Chinese samples.

MEASURES OF MATHEMATICAL PERFORMANCE

As mentioned earlier, an ideal cross-national study would involve an assessment of students' mathematical performance across a range of different, complementary test formats. Thus, this cross-national study used different formats of tests to capture a range of mathematical performances. In particular, the testing instrument used in this study consists of three sets of mathematical items: computation tasks that measure computation skills, component questions that measure simple problem-solving skills, and open-ended problems that measure complex problem-solving skills. The use of computation tasks, component questions, and open-ended problems allows for assessing a range of mathematical performance of the U.S. and Chinese students.

Computation Tasks

The first set of 20 multiple-choice items consists of arithmetic computation items. The items involve different operations (addition, subtraction, multiplication, and division) on different types of numbers (whole numbers, decimals, and fractions) and allow for examining U.S. and Chinese students' computation skills. Appendix A shows the English versions of the 20 computation items.[4]

Most (14 out of 20) of the computation items were selected from the fifth-grade computation test designed by Stigler et al. (1990) for their study of the mathematical knowledge of Japanese, Chinese, and U.S. elementary school children. Item selection was based on two criteria: (a) degree of difficulty and (b) embedded operations. Stigler designed the computation test for fifth-grade students. In this study, it was used with sixth-grade students. It is reasonable to assume that the sixth-grade U.S. students in this study are more competent than the fifth-grade U.S. students in the Stigler study. Thus, the 14 items chosen were those found to be difficult for U.S. fifth-grade students. In particular, on the basis of the performance of the fifth-grade students in Stigler's study, the difficulty index for 3 of the items is about .45 and the difficulty index for the other 11 items is less than .25. The remaining 6 items were developed by the researcher to provide balance for the different operations and the different types of numbers. The difficulty levels for these 6 developed items were estimated to be similar to the 14 selected items from Stigler.

Component Questions

The second set of tasks consists of 18 multiple-choice component questions that were designed to assess students' word-problem-solving component processes. The components are translation, integration, and planning (six items for each component). These questions were designed according to the components of solving simple word problems formulated by Mayer (1987). Thus, this set of questions facilitated examination of students' simple problem-solving skills. Appendix B shows the English versions of the 18 component questions.

These component questions were adapted from Mayer et al. (1991), who used these questions to compare U.S. and Japanese fifth-grade students' performance. It was reported in that study that the mean number of items that Japanese fifth-graders answered correctly was 11.2 (out of 18), and for U.S. fifth-graders it was 7.6 (out of 18). Therefore, this set of questions should be of appropriate difficulty for the U.S. and Chinese sixth-graders in this study.

Open-Ended Problems

The third set of tasks consists of seven open-ended problems that were designed to assess students' complex problem-solving skills. They are cognitively complex in at least the following ways: (a) There is not a routine that a student can follow to solve each open-ended problem; (b) the problems are in an open-ended format, where a student is asked not only to provide his or her own answer but also to explain the thinking process that yields the answer; (c) there are multiple approaches to solving each problem, and some even allow multiple correct answers; and (d) each problem has multiple conditions.

---

[4]The Chinese version of computation tasks, component questions, and open-ended problems are available from the author.

Open-ended problems allow students to construct solutions; provide a visible record of their solution processes; and allow for multiple representations, strategies, and solution justifications. Thus, these problems allow for examining U.S. and Chinese students' thinking and reasoning. The open-ended problems involve a variety of important content areas, such as number sense, pattern recognition, number theory, prealgebra, ratio and proportion, estimation, and statistics. The seven problems were adapted from the QUASAR Cognitive Assessment Instrument (QCAI) (Lane, 1993), with a minor revision for some of them. Appendix C shows five of the seven open-ended problems.

The first open-ended problem is a division-with-remainder story problem. Hereafter, this problem is called the Division Problem (OE-1). This task assesses a student's proficiency in choosing an appropriate strategy to solve a problem and is used later in making sense of her or his computational result by mapping it back to the problem situation. The Division Problem provides an interesting context in which to consider student mathematical thinking. In solving a division story problem involving a remainder, one not only needs to apply and execute division computations correctly but also to interpret correctly the computational result with respect to the given story situation. An earlier study by Cai and Silver (1995) showed that Chinese fifth- and sixth-grade students and U.S. sixth-, seventh-, and eighth-grade students have similar sense-making difficulties in solving this kind of division problem.

The second open-ended problem involves estimation and is directly taken from the QCAI. Hereafter, this problem is called the Estimation Problem (OE-2). The problem assesses student proficiency in estimating the area of an irregular shape. Estimation is a very important skill in mathematics. NCTM (1989) suggests that "the curriculum should include estimation so that students can explore estimation strategies and apply estimation in working with quantities, measurement, computation, and problem solving" (p. 36). In this task, students are asked to provide a reasonable estimate and then explain how they obtained their estimate. The irregular shape is placed on a grid. Solving the problem requires that students transform the irregular shape into a regular shape or use compensation in accounting for the partially shaded squares. Multiple estimation strategies and representations can be used to solve the problem.

The third problem involves the arithmetic mean and is taken from the QCAI with a minor modification (i.e., the numbers involved in this problem are slightly larger). Hereafter, it is called the Average Problem (OE-3). The problem is designed to assess students' understanding of the concept of average, one of the most important concepts in mathematics and a useful concept in everyday life. In particular, this task requires students to find the missing number when three numbers and the average of the three numbers and the missing number are presented in a picture. Students can use different solution strategies to determine the answer including trial and error, graphical representations, arithmetic calculations, or algebraic equations. In solving this Average Problem, students cannot simply manipulate numbers to find the answer; they must understand the average concept in order to solve the problem.

The fourth open-ended problem is a number-theory problem that was taken from the QCAI. The problem assesses students' number sense and the inability to use basic concepts of number theory to solve a problem. Hereafter, it is called the Number Theory Problem (OE-4). To solve this problem, students must find an unknown number that satisfies several conditions in a story context. Specifically, the student must find the total number of blocks in a set, given that one block remains when the set is partitioned into groups of size 2, size 3, or size 4. Thus, a correct numerical answer should have a remainder of 1 when divided by 2, 3, or 4. An implied condition of the problem is that the same set of blocks is partitioned each time. Several problem-solving strategies can be used to solve this problem, including the use of common multiples and the interpretation of remainders from division computations. Also, students can represent the problem in different ways: pictures, words, or mathematical expressions. The number 1 and any multiple of 12 plus 1 are correct answers (i.e., $1 + 12n$, for $n = 0$, 1, 2..., are correct answers).

The fifth problem is a modified version of a QCAI pattern problem. In the modified version of this problem, students are asked to draw both the fifth and seventh figures and describe how they found the seventh figure. Hereafter, this problem is called the Pattern Problem (OE-5). In the original version, students are only asked to draw the fifth figure and describe how they found it. This task assesses a student's reasoning ability (i.e., to identify underlying regularities in a figural pattern). It requires the student to use regularities to extend a pattern and to communicate the regularities effectively. More specifically, the pattern consists of a sequence of figures, each composed of a number of dots arranged in the shape of a trapezoid (the first figure is a triangle, which can be regarded as a special case of a trapezoid). The first four figures are given; the student is asked to draw the fifth and seventh figures and also to describe how he or she knew what the seventh figure would look like. In order to correctly find the seventh figure, the student needs to focus on regularities with respect to both the number of dots in each figure and the shape of the figures. This problem can be solved by using different strategies and representations.

The sixth is a ratio and proportion problem taken directly from the QCAI. The task assesses students' problem-solving skills in a map-reading context that involves ratio and proportion. Hereafter, it is called the Ratio and Proportion Problem (OE-6). Students need to demonstrate their understanding of a proportional relationship involving a map scale and a real distance scale by determining a missing value. Solving this problem requires students to identify the appropriate proportional relationships among the given values and a missing value. Multiple solution strategies and representations are possible in solving this problem.

The seventh task is a prealgebra problem that is taken from the QCAI. Hereafter, it is called the Prealgebra Problem (OE-7). This problem assesses students' understanding of incremental rates and their problem-solving skills involving informal algebraic thinking such as variables and relationships. Students are given incremental rates of earnings on a per-day basis for two persons and asked to determine the number of days each person worked at which the incremental rates will yield the same quantity. To solve this problem, students need to focus on two relationships: (a) The rate times the number of days is equal to the total earning and (b) the total earning for the first person is equal to that for the second person. Once the students have found one set of correct answers (i.e., the number of days each person has worked so that they have earned the same amount of money), they were told there were multiple correct answers and they were asked to try to find another answer to the problem. This problem has infinitely many correct answers, and the students may exhibit any of them. Several solution strategies may be used by the students, such as listing of multiples, repeated addition, tables or pictorial representations, guess and check, and equations.

Opportunity to Learn

At the time of the data collection, each mathematics teacher whose students participated in this study was asked to indicate whether his or her sixth-grade students had been taught enough information to correctly answer each of the computation tasks, component questions, and open-ended problems. The mathematics teachers were also asked to indicate whether their students were familiar with the task format (multiple-choice or open-ended) in each test booklet. Opportunity to learn (OTL) data are useful in interpreting performance differences between U.S. and Chinese students (Leinhardt, 1982). Appendix D shows the questionnaire for the booklet of open-ended problems. The questionnaires for the booklets of computation tasks and component questions are similar to that for the booklet of open-ended problems.

INSTRUMENT VALIDATION

Reliability and Validity

The computation tasks and component questions were adapted from previous cross-national studies (Mayer et al., 1991; Stigler et al., 1990) involving U.S., Chinese, and Japanese students. In this study, the reliability estimates (KR-20) for the computation tasks are .65 for the Chinese students and .77 for the U.S. students. The reliability estimates (KR-20) for the component questions are .81 for the Chinese students and .79 for the U.S. students.

The open-ended problems were adapted from the QCAI, which were validated on the basis of data from U.S. students by Lane, Stone, Ankenmann, and Liu (1994). Because these problems were used with different samples and have been modified from the original instrument, it is useful to provide additional reliability information. In this study, the Cronbach alpha coefficients (Cronbach, 1951) are .69 for the Chinese students and .77 for the U.S. students.

Chinese translation

The translation of the computation tasks into Chinese was straightforward, but Chinese language versions of the component questions and the open-ended problems were not previously created in prior research. Therefore, these two sets of tasks and their directions for test administration were translated into Chinese as part of this study. There are at least three concerns in language translation of a test: (a) Are the resultant Chinese wordings of tasks culturally appropriate and age-appropriate? (b) Do U.S. and Chinese students interpret tasks in the same way? (c) Does the Chinese version of the tasks assess the same content and the cognitive processes they are designed to measure in the English version? In this study, two methods were used to address these concerns and to ensure translation equivalence: English back-translation and pilot studies.

In the translation to Chinese, personal names, object names, terminology, and contexts were changed into appropriate words for Chinese students; this should not affect the mathematical difficulty of the tasks. For example, the English version of the Estimation Problem used square yards as the area unit. Because Chinese students are not familiar with square yards, it was changed to square kilometers in the Chinese translation. The context for the English version of some problems was also changed to appropriate contexts for Chinese students. For example, because Chinese students generally go home to have lunch and need not buy lunch at school, the context in the following component question is not appropriate for Chinese students, if direct translation of the context is used: "Lucia had $3 for lunch. She bought a sandwich for $.95, an apple for $.20, and a milk for $.45. How much money did she spend?" Chinese students do not eat sandwiches, apples, and milk for lunch. Thus, the context was changed: "Liu Wei has 3 Yuan. He bought a book for .95 Yuan, a pencil for .20 Yuan, and a notebook for .45 Yuan. How much money did he spend?"

One change introduced in the process of translation was the word "pattern," and this change might be more significant. The appropriate translation to Chinese for the English word "pattern" in the Pattern Problem is the word "model" or "sequence." But "model" or "sequence" was thought to be too advanced for Chinese sixth graders. Therefore, "pattern" was translated into Chinese as "a group of figures with a certain rule" in order to capture the meaning of the word. Pilot data from 186 fifth- and sixth-grade Chinese students showed that the Chinese translation of the meaning for "pattern" was appropriate for Chinese students and evoked behaviors similar to those of U.S. students (Cai & Silver, 1994). However, it seems that the directions for the Pattern Problem are more explicit in the Chinese version than in the English version.

As stated earlier, the use of a test in a cultural group other than the one for which it was originally designed has often led to sharp controversies (Berry et al., 1992). Therefore, it was necessary in this study to ensure the equivalence of the two language versions of the tests. In order to do so, two persons literate in both Chinese and English contributed to the translation tasks. The first person translated the English into Chinese, and then the second person independently translated the Chinese back into English. The English back-translation and the original English were consistent except for the intentional changes involving specific words like personal names, object names, contexts, and terminology.

After the back-translation, the Chinese version of the test items was edited by a professional Chinese school textbook editor to smooth sentences and ensure comprehension by students.

Pilot Studies

Three pilot studies have been conducted to examine (a) if the Chinese wordings of tasks were culturally appropriate and age-appropriate; (b) if U.S. and Chinese students interpreted tasks in the same way; and (c) if the Chinese version of tasks assessed the content and the cognitive processes they were designed to measure in the English version.

In the first pilot study, two of the open-ended problems (the Pattern Problem and the Division Problem) were piloted with 186 fifth- and sixth-grade Chinese students. The results from the pilot data showed that the problems were appropriate for the Chinese students (Cai & Silver, 1994). In particular, the wordings in Chinese were culture- and age-appropriate and appeared to assess the content and the cognitive processes they were designed to measure in the English version.

The second pilot study was conducted using the Chinese version of three sets of mathematical problems with an 11-year-old Chinese student (fifth grader) living in the United States. The student was asked to read each problem aloud and then to solve it. Because the purpose of the study was to see if the Chinese wording of each task was clear and understandable, once the student showed evidence that he was able to solve an open-ended problem, he was asked to go on to the next problem. This pilot study suggested that the wordings and directions in Chinese were clear. The student interpreted the tasks and directions in the way that was expected.

In the third pilot study, the open-ended problems were piloted with 21 sixth-grade students who attend a U.S. private school. Analyses of the pilot data showed that the task wordings were age-appropriate and that they assessed the content and the cognitive processes they were designed to measure. Two of the problems (the Number Theory Problem and the Average Problem) were found to be difficult. The rest of them were either easy to solve or at an intermediate level of difficulty. The mathematics teacher of the students in the pilot study indicated that the content embedded in each of the open-ended problems was important and that her sixth graders were expected to know it.

In summary, these three studies suggest that (a) the Chinese wordings of the tasks are culture- and age-appropriate; (b) U.S. and Chinese students interpret the tasks in the same way; and (c) the Chinese version of the tasks assess the content and the cognitive processes they were designed to measure in their English versions.

TESTING ADMINISTRATION

The computation tasks, component questions, and open-ended problems were presented in three different booklets. The computation booklet and the component booklet were administered in one class period. Students were allowed 20 minutes for the computation booklet and 15 minutes for the component booklet. For each of the computation tasks and component questions, students were asked to choose the correct answer from the four given choices. On another day, students were allowed 40 minutes to complete the open-ended problem booklet. For each of the open-ended problems, students were asked to produce answer(s) and to show how they obtained their answer(s).

Test booklets were administered by the students' regular classroom mathematics teacher. Administration directions were provided by the researcher in a letter with detailed description of the administration procedures sent to each sixth-grade mathematics teacher before the data collection.

DATA CODING

Data Coding for Responses to the Computation Tasks

Each student response to a computation task was coded correct or incorrect. Item analysis was conducted to determine patterns of procedural errors in incorrect responses. If a student omitted an item, the student response on the item was coded as incorrect.

Data Coding for Responses to the Component Questions

Each student response to the component questions was coded correct or incorrect. For incorrect responses, item analysis was used to examine the mathematical errors that occurred in solving these questions. If a student omitted an item, the student response on the item was coded as incorrect.

Students were asked to select their answers on the booklets for the computation tasks and component questions. Student answers were transferred to machine readable answer sheets. The accuracy of the transformation was checked by randomly selecting about 50 Chinese and 50 U.S. student tests. The accuracy of the transformation was over 99%.

Data Coding for Responses to the Open-Ended Problems

Each response to an open-ended problem was scored according to two analysis schemes: a holistic scoring scheme (or quantitative analysis) and a cognitive analysis scheme (or qualitative analysis) (Cai et al., in press-a; Cai et al., in press-b; Lane, 1993; Magone et al., 1994; Silver & Cai, 1993). In the quantitative analysis, each student response was assigned a numerical score by the researcher. In the qualitative analysis, each student response was examined by the researcher in detail in terms of the cognitive aspects of processing and solving the problem. A more elaborate description of the quantitative and qualitative analyses follows.

Quantitative analysis. A quantitative analysis of U.S. and Chinese student responses to the open-ended problems was conducted using the focused holistic procedure developed earlier for use in scoring students' responses to the QCAI (e.g., Lane, 1993; Silver & Lane, 1993). The QCAI holistic scoring procedure was accomplished by first developing a general scoring rubric. The general scoring rubric reflects three interrelated performance dimensions: mathematical conceptual and procedural knowledge, strategic knowledge, and communication. Then a specific rubric was developed for each task. The criterion specified at each score level for each specific rubric is guided both by the general rubric and by an examination of the responses produced by students. Using the specific rubrics, each student response was scored at each of five levels (0–4).

In general, to receive a score of 4, a student's explanation or solution process must show a correct and complete understanding of the problem. At score level 3, students' explanations or solution processes would basically need to be correct and complete, except for a minor error, omission, or ambiguity. To receive a score of 2, the explanation or solution process should show some understanding of the problem but would otherwise be incomplete. If a student's explanation shows a limited understanding of the problem, it would be scored as 1. If a student's answer and explanation show no understanding of the problem, the response would receive a score of 0, or if a student omitted the problem, the response would be scored as 0.

Qualitative analysis. In addition to examining the required numerical answer or drawing for each open-ended problem, a qualitative analysis of each response to the open-ended problems focused on four critical cognitive aspects: solution strategies, mathematical errors, mathematical justifications, and modes of representation. The rationale for examining these cognitive aspects in this study includes the fact that these aspects have been identified as important and significant dimensions in cognitive psychology in general (e.g., Royer et al., 1993) and in mathematical problem solving in particular (e.g., Charles & Silver, 1989; Silver, 1987). On the basis of these four aspects, a specific qualitative coding scheme for each problem was developed. In a few instances (e.g., the Average Problem, the Number Theory Problem, the Pattern Problem, and the Division Problem) the scheme already existed (Cai, 1995a; Cai et al., in press-b; Cai & Silver, 1995; Magone et al., 1994).

It should be noted that quantitative and qualitative analyses are not only interrelated but also complementary. The qualitative analysis of student responses was conducted with respect to solution strategies, mathematical errors, mathematical justification, and modes of representation. These categories correspond to the three dimensions (mathematical conceptual and procedural knowledge, strategic knowledge, and communication) in the quantitative analysis. The modes-of-representation and mathematical-justification categories relate to the dimension of mathematical communication; the solution-strategies category relates to the dimension of strategic knowledge; and the mathematical-errors category relates to the dimension of mathematical knowledge.

In the quantitative analysis, each student response to an open-ended problem was assigned a score based on a set of well-specified criteria of mathematical problem solving and reasoning. However, this procedure may conceal some aspects of students' performance. For example, different students can use different strategies to obtain a particular score level, and thus differences in strategies will not be evident using the qualitative analysis. Similarly, students may exhibit different mathematical misconceptions at a particular score level. To complement the quantitative analysis, the qualitative analysis provided a more detailed description of the kinds of strategies associated with high-level scores and the kinds of errors associated with lower-level responses.

Interrater agreement. In order to ensure a high reliability of coding student responses to open-ended problems, a graduate student literate in both Chinese and English was used as a second rater to check interrater reliability. Because of time and resource limitations, the coding reliability check focused on only two problems: the Average Problem (OE-3) and the Number Theory Problem (OE-4). The two raters independently coded about 50 U.S. and 50 Chinese student responses both quantitatively and qualitatively. Interrater agreement for the quantitative analysis was 89% for the Number Theory problem and 84% for the Average Problem. Interrater agreement for the qualitative analysis ranged from 86% to 93% for the Number Theory problem and 90% to 98% for the Average Problem. Because there were high agreement rates, results were reported only on the basis of the researcher's analyses.

Another issue associated with reliability coding is the possibility of changes in coding over time. Since the quantitative scoring of students' responses to the seven open-ended problems lasted more than 2 months, one concern was whether a person interprets student responses in the same way over a 2-month period. To this end, the researcher randomly selected about 30 Chinese and 20 U.S. students' responses for each of the open-ended problems and rescored them after all scoring was completed. The overall agreement was 94%.

DATA ANALYSIS

A variety of data analyses were performed on the student response data from the three sets of mathematical tasks. For convenience, each research question was first stated, and then the analysis related to answering it was summarized. It should be noted that more than one statistical analysis was conducted for each measure of mathematical performance (computation skill, simple problem solving, and complex problem solving). Moreover, more than one statistical analysis was conducted for each of the open-ended problems. To guard against accumulated errors across multiple statistical tests, the significance level was set at $\alpha = .01$ in this study. The OTL data from mathematics teachers were analyzed and taken into account when the results of other analyses were interpreted.

1. What are the similarities and differences in mathematical performance between a group of U.S. students and a comparable group of Chinese students as measured by three types of tasks: performing computation tasks, assessing the component processes (translation, integration, and planning) in word problem-solving, and solving a set of open-ended problems?

Using students' quantitative scores on three measures of mathematical performance, a one-way MANOVA (multivariate analysis of variance) was conducted to determine whether U.S. and Chinese students differ in their mathematics performance as measured by the computation tasks, component questions, and open-ended problems. In the MANOVA, the independent variable was nationality (U.S. or Chinese), and the three dependent variables were the types of mathematical performance, as measured by the computation tasks, component questions, and the open-ended problems.

A significant difference was found on the overall MANOVA; post hoc analyses using regular *t*-tests with adjusted significant levels of $\alpha/3$ (Stevens, 1992) were used to test—(1) if U.S. and Chinese students differ in their success at performing the computation tasks; (2) if U.S. and Chinese students differ in their success at performing the component questions involving translation, integration, and planning; (3) if U.S. and Chinese students differ in their success at performing the open-ended mathematical problems.

Since the component questions measure student processes involving the translation, integration, and planning of word problem solving, whether or not U.S. and Chinese students differ in any of these three processes was also determined. A one-way MANOVA was conducted on students' scores on three component processes. The independent variable was nationality (U.S. and Chinese), and the three dependent variables were the types of mathematical performance as measured by component questions involving translation, integration, and planning. A significant difference was found on the overall MANOVA; post hoc analyses using regular *t*-tests with adjusted significance levels of $\alpha/3$ were conducted to determine the specific component that contributes to the difference.

2. What is the nature of the cognitive similarities and differences between a group of U.S. students and a comparable group of Chinese students in their performance in their use of strategies, their use of modes of representation, the errors they make, and the mathematical justifications they give in solving a set of open-ended mathematical problems?

Qualitative analyses that compared U.S. and Chinese student performance on each open-ended problem were conducted to identify cognitive similarities and differences between U.S. and Chinese students as they solve open-ended problems. This part was descriptive in nature, although some statistical analyses were used.

3. When levels of computational performance are matched, how does a subgroup of U.S. students differ from a comparable subgroup of Chinese students in their success in answering questions assessing word problem-solving component processes and solving a set of open-ended problems?

U.S. and Chinese students were matched on the basis of their performance on computation tasks, and then their performance on the component questions and the open-ended problems was compared and analyzed. On the basis of their performance on the computation tasks, several blocks of U.S. and Chinese students were selected. The selections were made in such a way that a similar number of U.S. and Chinese students were selected from each block and that U.S. and Chinese students had almost identical mean scores on the computation tasks in each block. Then on the basis of the block design, U.S. and Chinese student performance on the component questions and open-ended problems was compared. Hereafter, this analysis is referred to as the "blocking analysis." The detailed procedure of the block analysis was described in the next chapter when relevant results were presented.

4. RESULTS

The results of this study are reported in three sections: the results of the quantitative analyses, the results of the qualitative analyses, and the results of the relative comparisons between U.S. and Chinese student performance. The results of the quantitative analyses address the first research question; that is, what are the similarities and differences between U.S. and Chinese students in their mathematical performance as measured by computation tasks, component questions, and open-ended problems? The results of the qualitative analyses address the second research question; that is, what is the nature of the cognitive similarities and differences between U.S. and Chinese students in solving open-ended problems? In particular, the qualitative results reveals whether U.S. and Chinese students exhibited similarities and differences in their solution strategies, modes of representation, and mathematical errors. The last section addresses the third research question by reporting a comparison of U.S. and Chinese student performance on component questions and open-ended problems after adjusting for their performance on computation tasks.

RESULTS OF THE QUANTITATIVE ANALYSIS

General Quantitative Results

Figure 3 shows the distribution of scores for U.S. and Chinese students on computation tasks, component questions, and open-ended problems. On each graph, the horizontal scale shows the students' score level, and the vertical scale shows the percentage of students answering correctly. The maximum score is 20 for the computation tasks, 18 for the component questions, and 28 for the open-ended problems.

As Figure 3 indicates, the largest difference between U.S. and Chinese students' scores is in their performance on the computation tasks. There is also a difference between the distributions for the two groups on the component questions, but the distributions are quite similar for the open-ended problems. The distributions indicate that a larger percentage of Chinese students than U.S. students achieved high scores on the computation tasks; that is, Chinese students outperformed U.S. students on the computation tasks. This finding is consistent with those reported in many cross-national studies (e.g., Stevenson et al., 1990).

Although the distribution for Chinese students is more negatively skewed than the distribution for U.S. students on the component questions, this difference between distributions is smaller than on the computation tasks. Therefore, a larger percentage of Chinese students achieved higher scores than U.S. students on the component questions, but the difference for the component questions is smaller than for the computation tasks.

Figure 3 shows that the distribution of U.S. students' scores on the open-ended problems is almost the same as that of Chinese students. Thus, U.S.

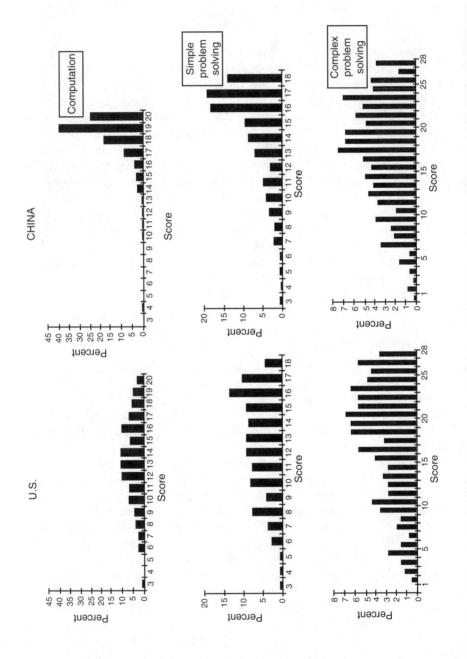

Figure 3. Distribution of scores for U.S. and Chinese students on computation tasks, component questions, and open-ended problems.

students performed similarly to Chinese students on the open-ended problems that were intended to measure complex problem-solving skills.

The mean scores of U.S. and Chinese students on computation tasks, component questions, and the open-ended problems are shown in Table 1. As was previously inferred from Figure 3, Chinese students had higher mean scores than U.S. students on the computation tasks and component questions. However, that was not true for the open-ended problems, where the mean score for U.S. students was greater than for Chinese students. A one-way multivariate analysis of variance (MANOVA) was used to test the significance of the difference between U.S. and Chinese students' performance across the three types of items. Nationality served as the independent variable, and the three types of test scores on computation tasks, component questions, and open-ended problems were the dependent variables. The analysis showed that there was a significant difference in mean performance between the U.S. and Chinese students among the three types of tasks ($F(3, 671) = 265.24$, $p < .001$). Thus, overall, Chinese students outperformed U.S. students on these tasks.

Table 1
Mean Scores of Chinese and U.S. Students on Computation Tasks, Component Questions, and Open-ended Problems

	Computation	Component	Open-ended
U.S.	13.46	12.64	18.13
(n = 250)	(3.61)[a]	(3.60)	(6.66)
China	18.45	14.43	17.65
(n = 425)	(1.74)	(3.37)	(6.24)
Possible Score	20.00	18.00	28.00

[a]Numbers in parentheses are the standard deviations.

A post hoc analysis was conducted to determine which mean differences on the three types of items were significant.[5] The results from the post-hoc comparisons using a regular t-test with an adjusted significance level of $\alpha/3$ showed significant differences between U.S. and Chinese students' performance on computation tasks and component questions, but not on open-ended problems. The mean score for Chinese students was significantly higher than that of U.S. students on the computation tasks ($t(673) = 24.14$, $p < .001$) and on the component questions ($t(673) = 6.48$, $p < .001$). However, U.S. and Chinese students had similar mean scores in solving open-ended problems, and there was no statistically significant difference in performance on these problems.

As a result of the different possible scores a student can obtain for each of the three types of items, no direct comparison of mean scores among the three types

---

[5]Post hoc analyses also showed that there were gender differences, favoring boys, for U.S. students on the computation tasks and the open-ended problems but not on the component questions (Cai, 1995b). There were no gender differences for the Chinese sample on any of the types of tasks.

of items can be made. For example, a 1-point difference on computation is out of a possible 18 points, whereas a 1-point difference on open-ended problems is out of a possible 28 points. In order to allow direct comparisons, mean scores for each item type were transformed to a percentage score by dividing the mean score by the maximum score for that type of item, hereafter referred to as the *percent mean score*. Figure 4 shows U.S. and Chinese students' percent mean scores on the three types of items. The greatest difference between U.S. and Chinese students' percent mean scores was on the computation tasks, where the Chinese students' mean score was 25 percentage points greater than the U.S. students' mean score. The Chinese students' mean score was 10 percentage points greater on the component questions, and 1 percentage point less on the open-ended problems.

The performance of Chinese students differed across the three types of items. Chinese students were most successful in solving the computation tasks, somewhat less successful on the component questions, and had the most difficulty with the open-ended problems. However, U.S. students had very similar performance levels for all three types of items, a performance level that was about the same as the Chinese on the open-ended problems but lower on the other two types of items.

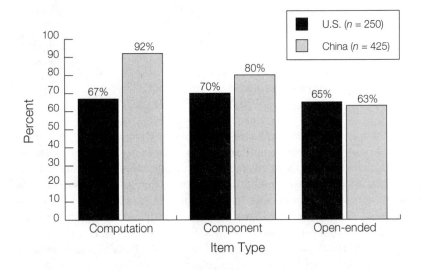

Figure 4. U.S. and Chinese students' percent mean scores on computation tasks, component questions, and open-ended problems.

The percentages of students who had perfect scores on each type of item provide another view of U.S. and Chinese students' performance. Figure 5 shows the percentages of U.S. and Chinese students who had perfect scores on each type of item. A larger percentage of Chinese students (24%) had correct answers for all the computation tasks than did U.S. students (4%) (x^2 (1, $N = 675$) = 50.10, $p < .0001$). Chi-square analysis also showed that a significantly larger percentage of Chinese students (14%) than U.S. students (5%) had correct answers for all the component questions (x^2 (1, $N = 675$) = 13.80, $p < .001$). However, the percentage of Chinese students (4%) who had perfect scores on the open-ended problems was the same as that of U.S. students (4%). The differential performance of the Chinese students across the three types of items compared to the stable performance of the U.S. students across these items is even more apparent from the results shown in Figure 5 than those in Figure 4. The percentage of the Chinese students receiving perfect scores dropped sharply from the computation to the component questions to the open-ended problems. The percentage of the U.S. students who obtained perfect scores on each of the three types of items remained constant.

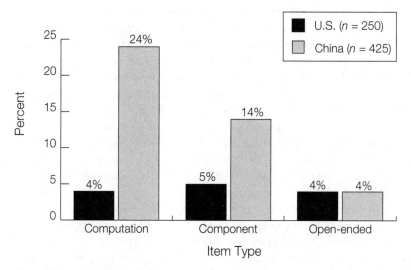

Figure 5. Percentages of U.S. and Chinese students who had perfect scores on three types of items.

A Detailed Examination of Performance on Computation Tasks

When asked about student readiness, all Chinese teachers (10) and most U.S. teachers (6 out of 9) responded that their students should have the necessary mathematical knowledge to solve all the computation tasks except the one involving addition with negative numbers. All U.S. and Chinese teachers reported they had not yet taught the mathematical knowledge necessary to solve this problem. Furthermore, three U.S. teachers responded that they were not sure that their students had the necessary mathematical knowledge to solve some computational tasks involving multiplication and division with fractions ($N = 5$ items).

A follow-up interview with these three teachers showed that their students had been taught multiplication and division with fractions in the previous term, but they had not reviewed the topic during the term in which the data were collected. All the U.S. and Chinese teachers felt that their students should be familiar with the format of the computational tasks. The results of teacher questionnaires suggested that both the U.S. and Chinese student samples had been taught the necessary mathematical knowledge to solve 19 of the 20 computation items. The 20th item involved an operation with negative numbers that should have been equally unfamiliar to both student samples, since no students had covered the topic. Therefore, the samples were matched on readiness, and the results in the previous section and this one are based on all 20 computation tasks.

Chinese students clearly outperformed U.S. students in solving the 20 computation tasks (mean of 18.45 correct for the Chinese and 13.46 for the U.S.). Superior computational performance can also be examined by determining the number of computation items for which over 90% of the students chose correct answers. There were 18 items where over 90% of Chinese students chose the correct answers, but only 3 items where over 90% of U.S. students chose the correct answer. Chinese students performed better than U.S. students on all computation items except the one addition problem involving negative numbers. On this item, 42% of Chinese students chose the correct answer, compared to 68% of U.S. students. Moreover, 14% (60 of 425) of Chinese students gave no response to this problem, but only 1% (3 of 250) of U.S. students gave no response.

Across all computation items, there was no single item for which every U.S. and every Chinese student chose the correct answer. Appendix E shows the results of the item analysis of the computation tasks in booklet order. U.S. students exhibited the most difficulty when trying to solve computation tasks involving fractions. Table 2 shows the mean success rates for U.S. and Chinese students when solving the seven computation tasks involving fractions and the thirteen computation tasks not involving fractions. Across the seven computation tasks involving fractions, the mean success rate of the U.S. students was 50%; this is significantly lower than the success rate of the Chinese students, 96% ($z = 14.16$, $p < .0001$). Although the mean success rate of Chinese students in solving nonfraction computation tasks (90%) was also significantly higher than that of U.S. students (77%) ($z = 4.59$, $p < .001$), the difference on the nonfraction computation

Table 2
Mean Success Rates of U.S. and Chinese Students' Solving Fraction and Nonfraction Computation Tasks

	Fraction computations	Nonfraction computations
U.S. ($n = 250$)	50%	77%
China ($n = 425$)	96%	90%

tasks (13 percentage points) was smaller than that on the fraction computations (46 percentage points). In fact, for U.S. students, the mean success rate on the nonfraction computation tasks was much higher than that on the fraction computation tasks, whereas the mean success rate for Chinese students on the fraction computation tasks was only slightly higher than that on the nonfraction computation tasks. Thus, Chinese student performance was more stable than U.S. student performance across fraction and nonfraction computation tasks. Poor U.S. student performance on the tasks involving fractions contributed substantially to their overall lower performance on the computation tasks as compared to Chinese students.

A detailed examination of U.S. student responses to the fraction computation tasks revealed their mathematical errors in solving these problems. For example, in solving

$$\frac{3}{8} \div 4 = ?$$

32% of the U.S. students picked $3/2$ as the answer, as compared to only 5% of the Chinese students. This incorrect answer most likely came from multiplying (rather than dividing) $3/8$ by 4. Similarly, for

$$\frac{5}{11} \div \frac{1}{9} = ?$$

40% of the U.S. students picked $5/99$ as the answer, as compared to only 2% of the Chinese students. The incorrect answer probably came from directly multiplying denominator by denominator and numerator by numerator (rather than $5/11 \times 9$).

In solving a multiplication task with one whole number and one fraction as factors, the common error for U.S. students was that they multiplied both the denominator and numerator of the fraction by the whole number. For example, given

$$6 \times \frac{4}{7} = ?$$

almost half of the U.S. students incorrectly chose $24/42$ as the answer. Only 1 out of 425 Chinese students picked $24/42$ as the answer. In solving

$$\frac{3}{4} - \frac{1}{6} = ?$$

about a quarter of the U.S. students answered $2/2$, a result that came from subtracting the smaller numerator from the larger numerator and subtracting the smaller denominator from the larger denominator—that is,

$$\frac{3}{4} - \frac{1}{6} = \frac{3-1}{6-4}.$$

Similarly, about 25% of the U.S. students found the sum of two fractions with unlike denominators by adding numerators and denominators. For example, in solving

$$\frac{5}{6} + \frac{3}{4} = ?$$

U.S. students selected $8/10$ as the correct answer.

U.S. students also had difficulty with place value when solving computational tasks involving decimals. For example, 60% of U.S. students were able to obtain the correct answer to the problem: $.08 \times 10 =$ __. One-fifth of the students chose .08 as the answer and about the same proportion chose 80 as the answer. For the problem: $36 \div .025 =$?, only 36% of the U.S. students selected the correct answer. Chinese students had little trouble in finding correct answers to items that involved multiplication and division of decimals.

A Detailed Examination of Performance on Component Questions

When asked if their students were familiar with the task format (i.e., select which numbers should be used to solve a problem, without solving the problem itself) in the component-question booklet, eight (of ten) Chinese and three (of nine) U.S. teachers responded that their students were not. However, all Chinese and U.S. teachers said that their students had been taught enough information to answer any of the component questions. Thus, the task format was quite new for most of the Chinese students, but they had been taught enough information to answer the component questions. One of the Chinese teachers put a note on the questionnaire that said, "We have never asked our students to do this kind of problem, but these problems are pretty simple." Therefore, the samples were similar in readiness, and the results reported are based on all 18 component questions.

As stated previously, the component questions were designed to measure students' simple problem-solving skills. In particular, the component questions measured students' performance involving the translation, integration, and planning component processes of solving word problems. As already reported, the aggregated mean score across the three types of component questions for Chinese students ($M = 14.43$) was significantly greater than for U.S. students ($M = 12.64$). A MANOVA was conducted to examine the overall difference between U.S. and Chinese students' performance on the component questions. It showed that, overall, Chinese students performed better than U.S. students on the component questions ($F(3, 673) = 30.38, p < .001$). This result is consistent with the analysis based on the aggregated mean scores of the three types of component questions. Therefore, Chinese students outperformed their U.S. counterparts in simple problem solving.

U.S. and Chinese student performance on each component was also examined separately. Table 3 shows the mean scores on the integration, translation, and planning component questions. Post hoc analyses using regular *t*-tests with adjusted significance levels of $\alpha/3$ showed that the overall difference between U.S. and Chinese students on the component questions resulted from differences on translation ($t(673) = 8.09, p < .001$) and planning ($t(673) = 6.22, p < .001$) but not integration ($t(673) = 1.28, p > .05$). This result suggests that Chinese students performed better than U.S. students on the six translation and the six planning component questions but that they performed equally well on the six integration problems. Caution is needed in interpreting the results related to each component process, because only six items were used to measure each component process.

Table 3
Mean Scores of U.S. and Chinese Students on Translation, Integration, and, Planning Component Questions

	Translation	Integration	Planning
U.S. (*n* = 250)	4.1	4.6	4.0
China (*n* = 425)	5.0	4.8	4.6
Possible Score	6.0	6.0	6.0

Appendix F shows the item analysis for the component questions in booklet order. For 16 of the 18 items, the most frequently chosen distracters by the U.S. and Chinese students were the same. This implies that, although Chinese students outperformed U.S. students on the component questions, Chinese and U.S. students made similar mathematical errors in solving the component questions.

On integration component questions (items 1, 4, 7, 10, 13, and 16) U.S. students did as well as Chinese students. The mean percent correct was 77% for U.S. students and 80% for Chinese students.

Of the planning component questions (items 2, 5, 8, 11, 14, and 17) the most difficult items for both U.S. and Chinese students were items 11 and 14. Low student performance on these items may be due to the design problems with these items. For example, the correct choice for item 11 (see Figure 6) is "a" (subtract, then multiply; i.e., 3 − 1 = 2, then 2 × 50 = 100). Since the subtraction (3 − 1) is so obvious, many students may not have viewed it as a necessary step in solving the problem. If so, they may have seen multiplication as the only operation that was needed. In fact, more than one-third of the U.S. and Chinese students chose "d" (multiply only) as the correct choice.

Similarly, for item 14 (see Figure 7) the correct choice is "b" (multiply, then divide; i.e., 2 × 20 = 40, then 40 ÷ 10 = 4). A considerable number of both U.S. and Chinese students, however, chose "c" (divide only) or "d" (multiply only) as their correct choice. Since 2 × 20 = 40 is a simple operation, some students may not have seen this as a necessary step in solving the problem, and therefore they chose "c" (divide only) as the answer. This problem can also be solved by using just multiplication. For example, 2 × 20 = 40, so they need 40 cookies. Because 1 box = 1 × 10 = 10 cookies, 1 box is not enough. Similarly, 2 boxes = 2 × 10 = 20, 3 boxes = 3 × 10 = 30, so 2 boxes or 3 boxes are still not enough. However, 4 boxes = 4 × 10 = 40 cookies. So 4 boxes are needed. Solving the problem in this way requires only multiplication. Therefore, choice "d" (multiply only) is a reasonable alternative answer.

The response patterns for U.S. and Chinese students were different on item 14. The most frequent incorrect answer for U.S. students (28%) was "d" (multiply only), but the most frequent incorrect answer for Chinese students (22%) was "c" (divide only). Thus, Chinese students seemed less likely than U.S. students to view the obvious operations as necessary steps. Further, U.S. students may have used the multiple-multiplication procedure for answering this question.

Which operations should you carry out to solve this problem?

If it costs 50 cents per hour to rent roller skates, what is the cost of using the skates from 1:00 p.m. to 3:00 p.m.?

     a.   subtract, then multiply
     b.   subtract, then divide
     c.   add, then divide
     d.   multiply only

Figure 6. Component item 11.

Which operations should you carry out to solve this problem?

You need to bring enough cookies so everyone at the class party can have 2 cookies each. There are 20 people at the party. Cookies come in boxes of 10 cookies each. How many boxes should you bring?

     a.   divide, then add
     b.   multiply, then divide
     c.   divide only
     d.   multiply only

Figure 7. Component item 14.

Among the six translation component questions (items 3, 6, 9, 12, 15, and 18), two involved an additive proposition (items 3 and 6). For example, "Ann and Rose have 20 books altogether" in item 3 is an additive proposition. The remaining four questions involved a relational proposition (items 9, 12, 15, and 18), as in "John has 5 more marbles than Pete" (item 9). U.S. students did as well as or better than Chinese students on the two translation questions involving an additive proposition. However, Chinese students outperformed U.S. students on all four translation questions involving a relational proposition. If the four translation questions involving a relational proposition were excluded in the analysis, the mean score of the remaining 14 items for U.S. students ($M = 10.4$) would be more closely aligned with that for Chinese students ($M = 11.2$), although the difference would still be statistically significant ($t (673) = 3.75, p < .001$). Therefore, the superior performance of Chinese students on the component questions was due, in part, to their higher performance on the translation questions involving a relational proposition.

The Chinese students' comparative ease in solving these latter questions may be partly related to differences in linguistic structures in Chinese and English.

In Chinese, the comparative nature in a relational proposition is explicitly shown whereas in English, the comparative nature is not explicitly shown in many cases. Consider again the example "John has 5 more marbles than Pete." This statement can be misinterpreted as an assignment proposition, since the comparative nature of the statement is only shown in the words "more ... than." In Chinese, "John has 5 more marbles than Pete" is read as "John-compare-Pete-more-5-marbles." The comparative nature is explicitly shown by the word "compare" (or " 比 " in Chinese, which is read as "bi") in the statement.

Chinese students outperformed the U.S. students on the translation component questions, but both groups made similar errors. The most frequent incorrect answers chosen by U.S. and Chinese students for the translation questions involving a relational proposition were the same and, interestingly, resulted from interpreting a relational proposition as an assignment proposition. For example, among the U.S. and Chinese students who chose an incorrect answer for item 9, the majority chose "b" (John's marbles + 5 = Pete's marbles) as the "correct" number sentence to represent the relationship in "John has 5 more marbles than Pete." This common error has been found in several other studies (e.g., Cocking & Mestre, 1988; Mayer, Lewis, & Hegarty, 1992). Therefore, although in the Chinese language a comparative nature is explicitly shown in a relational proposition, many Chinese students made errors by interpreting it as an assignment proposition. This also suggests the cognitive complexity involved in understanding a relational proposition by students.

A Detailed Examination of Performance on the Open-Ended Problems

When asked if their students were familiar with the task format for the open-ended problems, nine (of ten) Chinese and seven (of nine) U.S. teachers responded that they were. All Chinese and U.S. teachers said that their sixth-grade students had been taught enough information to answer the Division Problem (OE-1), the Number Theory Problem (OE-4), the Pattern Problem (OE-5), and the Ratio and Proportion Problem (OE-6). Five (of ten) Chinese and two (of nine) U.S. teachers responded that their students may not have been taught enough information to solve the Estimation Problem (OE-2). Two U.S. teachers and one Chinese teacher said their students may have not taught enough information to solve the Average Problem (OE-3). One U.S. teacher and two Chinese teachers responded that their sixth-grade students may not have been taught enough information to solve the prealgebra Problem (OE-7). Thus, most of the Chinese and U.S. students should have been familiar with the open-ended problem format and should have been taught enough information to solve most of the open-ended problems. In general, the students seemed to be close on readiness, and the results reported are based on all seven open-ended problems.

Each student response to an open-ended problem was scored using a five-point scale (0 – 4) with 0 = no understanding, 1 = beginning understanding, 2 = some understanding, 3 = nearly complete and correct understanding, and 4 = complete and correct understanding.

Appendix G shows the percentage distributions of scores for the U.S. and Chinese students for each open-ended problem. Overall, they performed equally well on the open-ended problems. The performance patterns on each of the open-ended problems were different. For some problems, the U.S. students did better than the Chinese students, and for other problems, the reverse was true.

Table 4 shows the mean scores of the U.S. and Chinese students on each of the open-ended problems. For three of the open-ended problems (OE-1, OE-4, and OE-7), U.S. students had greater mean scores than Chinese students, and for two other problems (OE-3 and OE-6), Chinese students outperformed U.S. students. For the remaining two problems (OE-2 and OE-5), U.S. and Chinese students had almost identical means. The Number Theory Problem (OE-4) was the most difficult one for both groups, and the Pattern Problem (OE-5) was the easiest for both. The greatest differences in performance were found for the Average Problem (OE-3) and the prealgebra Problem (OE-7). On the Average Problem, Chinese students outperformed U.S. students; however, on the prealgebra Problem, U.S. students outperformed Chinese students. It is reasonable to ask, "Why did U.S. students perform better than Chinese students on some problems, but not on others?" This question was addressed in the section where qualitative results based on detailed cognitive analyses of student responses were reported.

Table 4
Mean Scores of U.S. and Chinese Students on Each of the Open-Ended Problems

	OE-1* Division	OE-2 Estimation	OE-3* Average	OE-4 Number Theory	OE-5 Pattern	OE-6 Ratio and Proportion	OE-7* Prealgebra
U.S. ($n = 250$)	2.99	2.22	2.23	2.10	3.03	2.61	2.94
China ($n = 425$)	2.73	2.13	3.11	1.85	3.06	2.77	2.00

*For this problem, the difference in mean scores is statistically significant ($p < .01$).

Mean scores are one way of representing relative performance, and percentage distributions of scores give another view. Figure 8 shows the percentage distributions of U.S. and Chinese students receiving either of the two highest score levels (3 or 4). Results displayed on Figure 8 are consistent with mean performance, shown in Table 4. For example, as in Table 4, data in Figure 8 show that Chinese students outperformed U.S. students on the Average Problem (OE-3), and, conversely, U.S. students outperformed Chinese students on the prealgebra Problem (OE-7).

It is important when comparing performance levels to also examine the response rates. Table 5 shows the percentage distributions of U.S. and Chinese students who omitted one problem, two problems, or three or more problems and students who attempted every problem. A significantly larger percentage of U.S. students (93%) than Chinese students (63%) attempted all the open-ended problems ($x^2 (1, N = 675) = 73.49, p < .0001$). In other words, a larger percentage of Chinese students (37%) than U.S. students (7%) did not attempt at least one open-ended problem. Moreover, the percentage of Chinese students who omitted at least three problems was comparable to the percentage of U.S. students who omitted only one problem.

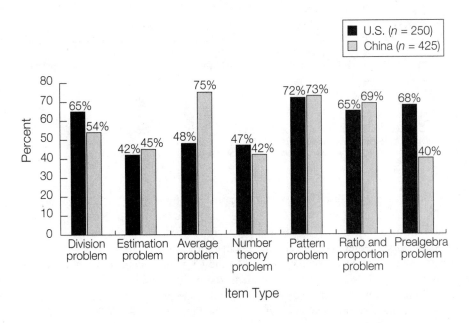

Figure 8. Percentage distributions of U.S. and Chinese students receiving the two highest score levels (3 or 4).

Table 5
Percentage Distributions of U.S. and Chinese Students Not Responding to Open-Ended Problems

	Omitted only one	Omitted only two	Omitted three or more	Responded to all problems
U.S. (n = 250)	5.2	1.2	.4	93.2
China (n = 425)	17.6	12.5	6.8	63.1

U.S. students not only had a higher overall response rate than Chinese students, but a larger percentage of U.S. students than Chinese students also attempted each of the open-ended problems. Table 6 shows the frequency distributions of U.S. and Chinese students who did not respond to each of the open-ended problems. Chi-square analyses were conducted on each of the open-ended problems, and the results indicated that the frequencies of U.S. students who did not respond to OE-2, OE-3, OE-4, OE-6, and OE-7 were significantly different from those of Chinese students (x^2 (1, $N = 675$) = 8.38 – 48.62, $p < .001$).

Table 6
Frequency and Percentage of U.S. and Chinese Students Not Responding to Each Open-Ended Problem

	OE-1 Division	OE-2* Estimation	OE-3* Average	OE-4* Number Theory	OE-5 Pattern	OE-6 Ratio and Proportion	OE-7* Prealgebra
U.S. ($n = 250$)	0 (0.0)[a]	2 (0.8)	1 (0.4)	7 (2.8)	0 (0.0)	4 (1.6)	8 (3.2)
China ($n = 425$)	1 (0.2)	59 (13.9)	18 (4.2)	82 (19.3)	6(1.4)	54 (12.7)	99 (23.3)

[a]The number in parentheses is the percent.
*For this problem, the difference in mean scores is statistically significant ($p < .01$).

Whereas only a few Chinese students did not attempt the Division Problem (OE-1) and Pattern Problem (OE-5), every U.S. student did. Interestingly, U.S. and Chinese students did almost equally well on these two problems. The Number Theory Problem (OE-4) and the Prealgebra Problem (OE-7) were the problems that the largest percentages of U.S. and Chinese students did not attempt. In particular, 19% of the Chinese students did not respond to the Number Theory Problem (OE-4), and 23% did not respond to the prealgebra Problem (OE-7). Whereas the Division Problem (OE-1) and the Pattern Problem (OE-5) gave students little difficulty, the Number Theory Problem (OE-4) and the Prealgebra Problem (OE-7) gave students the most difficulty. Chinese student scores on the open-ended problems in general, and their comparatively lower scores on the Number Theory Problem (OE-4) and the Prealgebra Problem (OE-7) in particular, reflect their high percentages of nonattempts, because a nonattempt to an open-ended problem was scored as zero.

Because a large percentage of Chinese students did not respond to the open-ended problems, it is informative to examine U.S. and Chinese students' performance when the problems not attempted are excluded from the analysis. In order to do that, an individual's score for the open-ended problems was calculated in the following way:

$$\text{Individual score} = \frac{\text{(sum of the scores on the attempted open-ended problems)}}{\text{(number of the open-ended problems attempted)}}$$

For example, if a student scored 3 on OE-1, 4 on OE-2, did not respond to OE-3, 2 on OE-4, 0 on OE-5, 4 on OE-6, and did not respond to OE-7, then the student's score on the open-ended problems is [(3 + 4 + 2 + 0 + 4) ÷ 5] = 2.6. When student performance on the open-ended problems was examined in this way, Chinese students had significantly larger mean scores (2.81) than U.S. students (2.62) (t (673) = 2.71, $p < .01$). Recall that there was no significant difference between U.S. and Chinese students' performance on the open-ended problems when the nonattempted responses were scored as 0 and included in the analysis. Thus, Chinese students had higher quality responses than U.S. students when only attempted responses were considered.

Table 7 shows the mean scores of U.S. and Chinese students on each of the open-ended problems when the nonattempt responses are excluded. The primary effect of excluding the nonattempts (cf., Table 4) is that the difference in performance on the Ratio and Proportion Problem (OE-6) is now statistically significant.

Table 7
Mean Scores of U.S. and Chinese Students on Each of the Open-Ended Problems, Excluding Non-attempt Responses

	OE-1* Division	OE-2 Estimation	OE-3* Average	OE-4 Number Theory	OE-5 Pattern	OE-6* Ratio and Proportion	OE-7* Prealgebra
U.S.	2.99	2.24	2.24	2.16	3.03	2.65	3.04
China	2.74	2.48	3.24	2.29	3.10	3.18	2.61

*For this problem, the difference in mean scores is statistically significant ($p < .01$).

RESULTS OF THE QUALITATIVE ANALYSES FOR THE OPEN-ENDED PROBLEMS

The qualitative analysis of student responses to the open-ended problems focused on four cognitive aspects of students' thinking and reasoning: solution strategies, mathematical errors, mathematical justifications, and mode of representations. On the basis of these four aspects, a specific qualitative coding scheme was developed for each problem. This section reports the results of the qualitative analyses of student responses to five of the open-ended problems (OE-1, OE-2, OE-3, OE-4, and OE-5). The reasons for choosing these five problems for detailed analysis were based on student performance considerations. U.S. students did significantly better than Chinese students on the Division Problem and the Prealgebra Problem, but because one-fourth of the Chinese students did not respond to the Prealgebra Problem, the Division Problem (OE-1) was chosen for the analysis. The Average Problem (OE-3) was chosen because it was the only item where Chinese students did significantly better than U.S. students. The Pattern Problem (OE-5) was chosen for qualitative analysis because it was the item for which both U.S. and Chinese students had the best performance among the seven open-ended problems. Because the Estimation Problem (OE-2) and the Number Theory Problem (OE-4) were the two most difficult items for both U.S. and Chinese students, both were included in the qualitative analysis. Therefore, the problems chosen for detailed analysis were the problems easiest or most difficult to solve for both U.S. and Chinese students, the problem for which U.S. students performed significantly better than Chinese students, and the problem for which Chinese students performed significantly better than U.S. students. The results are reported here by problem, with the analyses differing slightly across problems because of differences in the nature of the problems.

Qualitative Results for the Division Problem

The Division Problem (OE-1) assesses student proficiency in choosing an appropriate strategy to solve a problem and in later making sense of the computational result by mapping it back to the problem situation. Each student response to the Division Problem was coded on four distinct aspects: (a) solution process, (b) execution of procedures, (c) numerical answer, and (d) interpretation. The categorization scheme used in this analysis was adapted from Silver et al. (1993) and Cai and Silver (1995). The results are reported in three separate sections that address solution process and execution of procedures, numerical answer, and interpretation of the answer.

Solution process and execution of procedures. The solution process was defined as the set of procedures used by a student to obtain a numerical answer. The execution of procedures referred to the actions taken by the student in performing the procedure.

The majority of the U.S. and Chinese students recognized OE-1 as a problem that required a division computation. In particular, 98% of the Chinese and 93% of the U.S. students selected division. Beyond this, a few U.S. students used other appropriate procedures, such as repeated addition and repeated subtraction, to solve the problem. None of the Chinese students used appropriate procedures other than division. Small proportions of the U.S. (5%) and Chinese (2%) students used inappropriate procedures, such as multiplication, in trying to solve the problem.

With respect to the execution of procedures, a larger percentage of Chinese students (91%) than U.S. students (82%) executed the procedures correctly ($z = 3.39$, $p < .001$). Thus, although quite similar percentages of U.S. and Chinese students chose the correct procedures, more Chinese students correctly executed these procedures. The selection of appropriate procedures and then correct execution of the procedures is one of the two phases (computation and sense-making) in solving this division problem. The preceding results indicate that Chinese students outperformed U.S. students on the computation phase.

Numerical answer. A student's numerical answer was considered to be the number written by a student in the space provided for the answer to the Division Problem, "How many buses are needed?" Table 8 shows the numerical answers given by the U.S. and Chinese students and the percentage of the samples giving each answer. Chi-square analysis indicates that the distributions of Chinese and U.S. students who gave each type of numerical answer were significantly different (x^2 (5, $N = 675$) = 66.52, $p < .001$). In particular, a larger percentage of U.S. than Chinese students gave an answer of 13 ($z = 3.36$, $p < .001$), and a larger percentage of Chinese than U.S. students gave an answer of 12 ($z = 5.24$, $p <$

.001). A few Chinese students said that the answer could be either 12 or 13. For example, one student showed the division calculation "296 ÷ 24 = 12R8," and then wrote "(1) 8 people are left, another bus is needed for these 8 people. So the answer is 13. (2) Among 12 buses, you can choose 8 buses to hold the 8 people, each bus holds extra 1 person, therefore, you just need 12 buses." No U.S. students made such comments.

Table 8
Percentage Distributions of U.S. and Chinese Students' Numerical Answers for the Division Problem

	Percentage of Students	
Numerical answers	U.S. ($n = 250$)	China ($n = 425$)
13	65	52
12	11	28
12 and fractional remainder	0	6
12 and whole number remainder	10	2
12 and decimal remainder	2	4
Other answers or no answer provided	11	9
Total	99[a]	101[b]

[a]Percentage does not add to 100 because of rounding.
[b]Percentage does not add to 100 because of rounding.

Approximately 12% of both U.S. and Chinese student responses included a remainder. That remainder could have been expressed as either a fraction, a whole number, or a decimal. U.S. students tended to express the remainder as a whole number, whereas Chinese students tended to express the remainder as either a fraction or decimal. Recall that U.S. students had more difficulty solving computation tasks involving fractions and decimals than whole-number-related computations. Thus, U.S. students may have felt more comfortable expressing the remainder as a whole number.

Interpretations. Interpretations were students' written explanations of their solution processes. Three categories (appropriate interpretation, inappropriate interpretation, and no interpretation) were used by Silver et al. (1993) to classify interpretations. These categories for interpretation were extended by Cai and Silver (1995) to the following six categories: appropriate interpretation, direct rounding to 12, direct rounding to 13, procedural explanation, inappropriate interpretation, and no interpretation. In this study, the extended classification categories were used to code student responses to the Division Problem.

An interpretation was coded "appropriate" if a student explained that a whole number of buses was needed, because a fraction of a bus did not make sense, or provided some other reasonable justifications. An interpretation was coded "inappropriate" if a student explained the numerical answer by applying rounding rules or by giving evidence of confusion. A response was coded "direct rounding to 12," if the student did not provide an explanation of the solution but her or his work showed that the answer resulted from directly rounding the division computational result symbolically. For example, if a student wrote "296 ÷ 24 ≈ 12," then the student's response was coded as "direct rounding to 12." Similarly, if a student wrote "296 ÷ 24 ≈ 13," then the student's response was coded as "direct rounding to 13." If a student's explanation consisted only of the description of the execution of the steps of the algorithm, then the response was coded "procedural explanation."

Table 9 shows the percentages of students who provided various kinds of interpretations for their answers. A significantly larger percentage of U.S. students (47%) than Chinese students (29%) provided appropriate interpretations for their answers ($z = 4.72$, $p < .001$). Common examples of appropriate explanations for the answer of 13 consisted of written descriptions such as, "If you have 12 buses, there are 8 people left over. These 8 people cannot walk and they have to go. Therefore, you need one more bus." "Bus cannot be one-third. Bus should be either one bus or 2 buses. So answer should be 13." A few students provided appropriate explanations to support a final answer of 12. For example, one student wrote, "If you have 12 buses, there are 8 people left over. These 8 people can squeeze into the 12 buses. So they just need 12 buses."

Table 9
Percentage Distributions of U.S. and Chinese students' Interpretations in Solving the Division Problem

	Percentage of students	
Interpretation types	U.S. ($n = 250$)	China ($n = 425$)
Appropriate interpretation	47	29
Direct rounding to 13	4	13
Direct rounding to 12	0	19
No interpretation	47	38
Inappropriate interpretations	2	1
Total	100	100

A larger percentage of Chinese students than U.S. students directly rounded the computational result up (12.3, 12⅓, or 12 R 8) and provided an answer of 13, but these students did not justify why they rounded up the computational result ($z = 4.00$, $p < .001$). Moreover, about 20% of Chinese students directly rounded the computational result down (12.3, 12⅓, or 12 R 8) and provided an answer of 12, but no U.S. student did so.

Only a few U.S. and Chinese students explicitly applied the rounding rule in their explanations. An example of this kind of response is when a student provided the division calculation, got a computational result such as 12.3, then wrote, "The remainder is less than 5, so I just take the .3 away." About one-half of the U.S. students and one-third of the Chinese students did not provide an explanation for their answers. For example, one student showed division computational work, then directly recorded the computational result of 12⅓ as the answer without providing an explanation.

In solving the Division Problem, a student response was considered correct if the student recorded 13 as the answer with or without appropriate interpretation or an answer other than 13 with appropriate interpretation. On the basis of this criterion, more than one-half of both the U.S. and Chinese students provided correct solutions for the Division Problem. However, the percentage for U.S. students (68%) was significantly larger than for Chinese students (53%) ($z = 3.67$, $p < .001$). Thus, U.S. students outperformed the Chinese students on the sense-making phase of solving the Division Problem.

In summary, the analysis based on the quantitative scoring showed that the U.S. students had a significantly higher mean score than the Chinese students on the Division Problem (see Table 4). The results from the qualitative analysis suggest that U.S. students' superior performance on the problem was not due to their performance in the computation phase but rather to their work in the sense-making phase. In fact, a larger percentage of Chinese than U.S. students used appropriate procedures and executed them correctly, but the reverse was true for finding the correct numerical answer and appropriately interpreting their answers. A larger percentage of U.S. than Chinese students had the correct answer of 13 and a larger percentage of U.S. students provided appropriate interpretations for their answers with respect to the given bus situation. These results indicated that although Chinese students outperformed U.S. students on the computation phase, U.S. students outperformed Chinese students on the sense-making phase. This finding is consistent with results reported in previous studies (Cai & Silver, 1994, 1995) and the findings from this study that Chinese students outperformed U.S. students on computation tasks, but not on open-ended problems.

Qualitative Results for the Estimation Problem

The second open-ended problem analyzed was the Estimation Problem (OE-2) that assesses student proficiency in estimating the area of an irregular shape. Each student response for the Estimation Problem was coded with respect to three aspects: (a) estimate, (b) estimation strategy, and (c) mode of explanation. The results are reported in separate sections that follow.

Estimate. The number written by a student in the space provided for the estimation of the area of the island was considered to be a student's estimate. Student estimates were classified into four categories according to the accuracy of the estimate, as described below.

Good estimate	Between 55 and 61
Acceptable estimate	Between 50 and 54 or between 62 and 65
Poor estimate	Between 45 and 49 or between 66 and 74
Unreasonable estimate	Anything outside the ranges of good, acceptable, or poor estimates

Table 10 shows the percentages of U.S. and Chinese students who provided different types of estimates. Similar percentages of U.S. and Chinese students provided good estimates, acceptable estimates, and poor estimates. More than 40% of U.S. and Chinese students provided good estimates, about 20% of U.S.

and Chinese students provided acceptable estimates, and another 20% of each sample provided poor estimates. Poor estimates resulted from both underestimation and overestimation of the area for the island. For example, some students counted only the number of fully shaded squares, whereas others counted both fully and partially shaded squares. Ten percent of the U.S. students and 4% of the Chinese students provided unreasonable estimates. For some U.S. students, unreasonable estimates resulted from confusion between the concepts of area and perimeter. In fact, 11 U.S. students explicitly stated that they got their estimates by counting the number of squares around the island, but no Chinese students did so. Over 10% of Chinese students did not provide an estimate because they did not respond to the problem; only one U.S. student failed to provide an estimate.

Table 10
Percentage Distributions of U.S. and Chinese Students' Estimates

	Percentage of students	
Estimates	U.S. ($n = 250$)	China ($n = 425$)
Good estimate	42	45
Acceptable estimate	23	20
Poor estimate	23	18
Unreasonable estimate	12	4
No estimate	0	12
Total	100	99[a]

[a]Percentage does not add to 100 because of rounding.

Estimation strategy. Three common strategies were used to obtain the estimates:

Strategy 1 (Counting)—Students counted the completely shaded squares and partially shaded squares and combined them to get the estimate. For example, they counted 47 completely shaded squares and 22 partially shaded squares. Letting every two partially shaded squares count as one complete square, the 22 partially shaded squares contribute 11 units to the area. So the estimate in this example was $47 + 11 = 58$.

Strategy 2 (Reforming)—Students reformed the island into a rectangle with approximately the same area and used the area of the rectangle as an estimate for the area of the island.

Strategy 3 (Subtraction)—Students selected a figure (e.g., rectangle) that covered the island. Then they calculated the area of the enclosing figure and estimated the "extra" area that the figure covered. The estimate of the island was obtained by subtracting the "extra" from the area of the enclosing figure. For example, some students selected an 8-by-10 rectangle to cover the island. The area of the rectangle is 80 square yards. They estimated that the unshaded area in the rectangle was about 20 square yards. So the area of the island was estimated to be about 60 (80–20) square yards.

For the Estimation Problem and the rest of the open-ended problems, one form of the analysis dealt with the solution strategy that students employed. In some cases, the student's solution strategy was readily apparent from the explanation the student gave. The researcher of this study referred to these as "apparent strategies." In other cases, no strategy was apparent because the student's explanation was either incomplete or vague.

A larger percentage of U.S. students (88%) than Chinese students (75%) had readily apparent strategies ($z = 4.17$, $p < .01$). For those who had apparent strategies, chi-square analysis showed that there was no statistically significant difference between U.S. and Chinese students' frequencies of using various strategies. Figure 9 shows the percentage distributions of U.S. and Chinese students who used various estimation strategies. The majority of both groups used a counting strategy. Sixteen percent of the U.S. students used the reforming strategy, which is twice as many as the Chinese students (8%). A small proportion of both U.S. and Chinese students used the subtraction strategy.

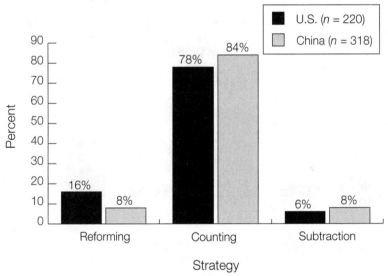

Figure 9. Percentage distributions of U.S. and Chinese students who used various estimation strategies for the Estimation Problem.

The reforming and the subtraction strategies appear to be more sophisticated than the counting strategy. It is interesting to note, however, that using the reforming and the subtraction strategies did not necessarily yield better estimates than using the counting strategy. In fact, only 23% of the U.S. and 31% of the Chinese students who used either the reforming or the subtraction strategy provided good estimates.

Mode of explanation. Only student responses that contained explanations (about 80% of the Chinese students and 95% of the U.S. students) were examined for the mode of explanation used. Mode of explanation was classified into one of three categories: verbal only, a combination of verbal and symbolic, or a combination of verbal and visual. The frequencies with which the students used various modes of explanation were different (x^2 (2, $N = 586$) = 8.71, $p < .01$). Figure 10 shows the percentage distributions of U.S. and Chinese students who used various modes of explanation. Every U.S. and Chinese student who provided an explanation used a verbal-related representation including verbal only, combination of verbal and symbolic, or combination of verbal and visual, in his or her solution process. Specifically, 74% of the U.S. students and 84% of the Chinese students used only verbal representation in their solution. A significantly larger percentage of U.S. students (51 of 243) than Chinese students (43 of 344) used visual-related representations in their estimations ($z = 3.78$, $p < .01$).

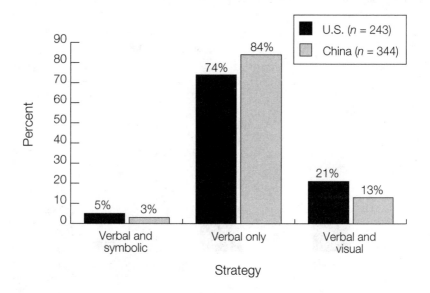

Figure 10. Percentage distributions of U.S. and Chinese students' modes of explanation for solving the Estimation Problem.

In summary, the quantitative analysis showed that the U.S. students had as high a mean score as the Chinese students on the Estimation Problem (cf. Table 4). The results from the qualitative analysis showed that similar percentages of U.S. and Chinese students provided each type of estimate. U.S. and Chinese students used three different estimation strategies with similar frequency, and the frequency of using each type of estimation strategy was also similar. Every U.S. and Chinese student that provided an explanation for their estimation processes used a verbal representation in their explanation. A larger percentage of U.S. students than Chinese students used a visual representation in their explanation.

Qualitative Results for the Number Theory Problem

The third open-ended problem analyzed was the Number Theory Problem (OE-4). Each student response was coded with respect to (a) numerical answer, (b) solution strategy, (c) mathematical error, and (d) mode of representation.

Numerical answer. As was indicated in the method section, the Number Theory Problem has more than one correct answer. In fact, the number 1 and any multiple of 12 plus 1 are correct answers (i.e., $1 + 12n$, for $n = 0, 1, 2,...$). Over one-half of the U.S. and Chinese students got correct answers. In particular, 56% (140 of 250) of the U.S. sample had correct answers, which is slightly more than for the Chinese sample, 54% (228 of 425).

The frequency distributions were significantly different for the U.S. and Chinese students who provided the correct answer of 13 and those who gave correct answers other than 13 (x^2 (1, $N = 368$) = 28.26, $p < .001$). As shown in Figure 11, a larger percentage of Chinese than U.S. students tended to provide correct answers other than 13. Correct answers other than 13 included 1, 25, 49, and so on. For those U.S. and Chinese students who provided correct answers other than 13, the majority of them had 25 as the correct answer. It is interesting to note that only two U.S. students and seven Chinese students provided more than one correct answer in their response. A few Chinese students provided the general form for the correct answers, but no U.S. student did so. Figure 12 displays several student responses with more than one correct answer and the general form of correct answers.

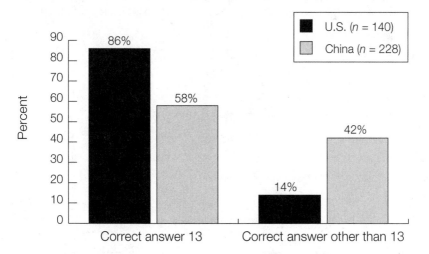

Figure 11. Percentage distributions of U.S. and Chinese students' correct answers for the Number Theory Problem.

Example 1

This U.S. student listed odd numbers, then crossed out the numbers that were not the answers (most likely by trial and error). The numbers remaining would be the correct answers.

Example 2

This Chinese student found the least common multiple of 2, 3, and 4, and then provided 13 and 25 as the correct answers. This student also wrote that any common multiple (of 2, 3, and 4) plus 1 would be a correct answer.

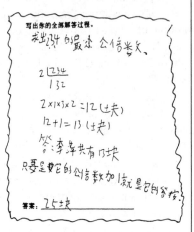

Example 3

This Chinese student checked that 13, 25, and 49 could be correct answers. Then the student said that if you doubled the previous answer, then minus 1, the obtained number would be a correct answer. This can be expressed by using mathematical notations: if n is a correct answer, then $2n - 1$ is also a correct answer (n is a positive integer).

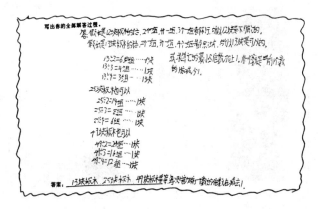

Figure 12. Sample student responses with more than one correct answer and the general form of correct answers.

Solution strategy. The Number Theory Problem evoked a variety of solution strategies. To solve the problem, an unknown number that satisfies several conditions must be found, and students in this study used at least eight different strategies to do it.

Strategy 1

The student found 12 as a common multiple of 2, 3, and 4 by direct computation ($2 \times 6 = 12$, $3 \times 4 = 12$, $4 \times 3 = 12$) and then added 1 to the common multiple (see Example 1 in Figure 13).

Strategy 2

The student found 24 as a common multiple of 2, 3, and 4 by direct computation ($2 \times 3 \times 4 = 24$) and then added 1 to the common multiple (see Example 2 in Figure 13).

Strategy 3

The student constructed three separate diagrams showing sets of blocks, each divisible into groups of 2, 3, or 4; attempted to make the sets all the same size; and then added 1 block to each of the sets. For example, the first set of blocks had 12 blocks that were grouped in groups of 2; the second set of blocks had 12 blocks that were grouped in groups of 3; and the third set of blocks had 12 blocks that were grouped in groups of 4. Then, 1 block was added to each set to obtain a total of 13 (see Example 3 in Figure 13).

Strategy 4

The student listed the multiples of 2, of 3, and of 4; identified the common multiple; and then added one. For example, the student did as follows:

2, 4, 6, 8, 10, 12, 14, 16,…

3, 6, 9, 12, 15, 18,…

4, 8, 12, 16, 20, 24,…

$12 + 1 = 13$, so the answer was 13.

Strategy 5

The student used "short division" to find the common multiple of 2, 3, and 4, and then added 1 to the common multiple to obtain a solution (see Example 4 in Figure 13).

Strategy 6

Other common multiple approaches.

Strategy 7

The student showed a number (e.g., 13) divided by 2, by 3, and by 4 in three separate long divisions, indicating that all the divisions had a remainder of 1. In this way, the student verified a correct answer by the divisions without showing how he or she obtained the dividend.

Strategy 8

The student systematically or randomly guessed numbers of blocks and checked if the guesses satisfied the conditions of the problem.

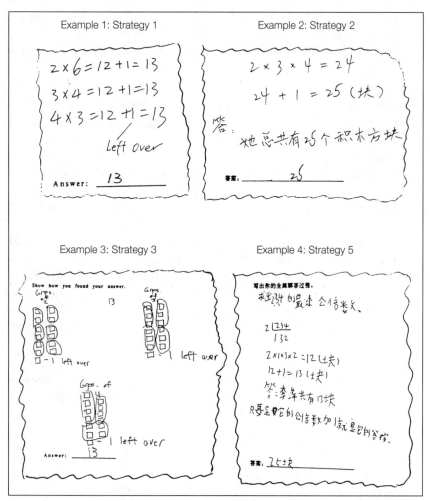

Figure 13. Sample student responses involving solution strategies in solving the Number Theory Problem.

Only a slightly larger percentage of U.S. students (52%) had apparent strategies than did Chinese students (45%), with the difference not statistically significant. However, for those U.S. and Chinese students with apparent strategies, the frequency distributions of strategies used were significantly different (x^2 (1, $N = 322$) = 154.81, $p < .001$). Table 11 shows the percentage distributions of the students with apparent strategies. Strategy 1 through strategy 6 involved common-multiple approaches, and strategies 7 and 8 used variations of a trial-and-error approach. For those U.S. and Chinese students with apparent strategies, a larger percentage of Chinese than U.S. students used common multiple approaches (strategy 1 through 6) to obtain solutions ($z = 4.78$, $p < .001$). A larger proportion of U.S. students than Chinese students used trial-and-error approaches (strategies 7 and 8).

Table 11
Percentage Distributions of U.S. and Chinese Students with Apparent Strategies for the Number Theory Problem

	Percentage of Students	
Strategy	U.S. (*n* = 129)	China (*n* = 193)
Strategy 1	19	2
Strategy 2	1	42
Strategy 3	22	1
Strategy 4	2	1
Strategy 5	2	32
Strategy 6	12	5
Strategy 7	29	13
Strategy 8	13	4
Total	100	100

Moreover, U.S. and Chinese students tended to use different variations of the common multiple approaches. For example, two common multiple approaches that were most frequently used by the Chinese students were strategies 2 and 5; those by the U.S. students were strategies 1 and 3. Strategies 1 and 2 are literally the same except for the computation use. The use of strategy 1 yields a correct answer of 13, and the use of strategy 2 yields a correct answer of 25. This may explain why a larger percentage of U.S. than Chinese students provided the answer of 13 and a larger percentage of Chinese than U.S. students provided the answer of 25. Twenty-eight U.S. students used strategy 3, but only two Chinese students used it. In strategy 3, visual representations were used to find a common multiple of 2, 3, and 4. As we see in the following section, overall, the U.S. students tended to use visual representations more frequently than the Chinese students. A considerable number of Chinese students used strategy 5, in which "short division" was used to find the common multiple (See Example 4 in Figure 13). In Chinese textbooks, the short-division approach is often introduced to find common multiples (Division of Mathematics of People's Education Press, 1987).

Mathematical errors. Excluded from the error analysis were students who did not respond to the problem, who responded to the problem but did not provide an explanation, or who got a score of 4. Thus, the responses of 155 Chinese (37%) and 135 U.S. (54%) students were included in this analysis.

Six types of errors were identified from students' solutions. Three error types reflected a failure to address one or more of the major conditions of the problem: that the whole set is partitioned into groups of 2, of 3, and of 4; that there is always a remainder of 1; or that the same set of blocks is partitioned each time. The six types of errors are described below.

1. Minor errors: Minor calculation error.
2. Omission of grouping(s): The blocks are grouped in 2s, 3s, or 4s, but one or two groupings are omitted.

3. Omission of the remainder in at least one grouping: Student solutions that omitted the remainder of 1 in at least one grouping.

4. Failure to conserve the total number of blocks: An implied condition of the task is that the same set of blocks is partitioned each time. (A student not meeting this condition might partition one set of blocks into groups of 2 and have 1 left over, partition another set into groups of 3 and have 1 left over, and so on, resulting in 3 different answers, none of which simultaneously satisfies all numerical constraints.)

5. Unjustified symbol manipulation: Students just picked some numbers from the task and worked with them in ways irrelevant to the problem context (for example, added the numbers together).

6. Other error, or errors cannot be identified.

Table 12 shows the percentage distributions of the errors made by the U.S. and Chinese students. A larger percentage of Chinese than U.S. students tended to make minor calculation errors. However, a larger percentage of U.S. than Chinese students tended to make errors that resulted from failing to satisfy one or more of the conditions (i.e., omission of grouping(s), omission of the remainder in at least one grouping, or failure to conserve the total number of blocks) ($z = 6.01$, $p <$ .001). In particular, 12% (16 of 135) of the U.S. students failed to conserve the total number of blocks, but no Chinese student did so. Interestingly, a considerable number of both U.S. and Chinese students manipulated the given numbers unreasonably in trying to solve the problem. Figure 14 shows a Chinese and a U.S. student response in which unjustified symbol manipulation occurred.

Table 12
Percentage Distributions of U.S. and Chinese Students Who Made Mathematical Errors in Solving the Number Theory Problem

	Percentage of Students	
Error types	U.S. ($n = 135$)	China ($n = 155$)
Minor error	4	15
Omission of grouping(s)	7	4
Omission of the remainder	19	5
Failure to conserve the total number of blocks	12	0
Unjustified symbol manipulation	33	42
Others	25	34
Total	100	100

Mode of representation. Only student responses that contained explanations (about 75% of Chinese students and 95% of U.S. students) were examined for the mode of representation used. In both samples, students who provided solution processes used various modes of representation. The modes of representation that students used were classified into three categories: verbal, symbolic, or visual. The response was coded as a verbal representation if a student mainly used written words to explain how he or she found the answer. The response

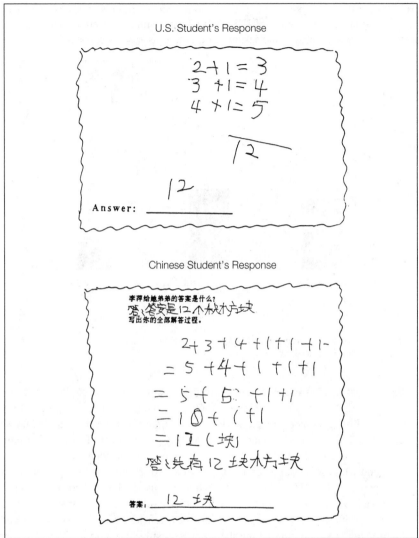

Figure 14. A Chinese and a U.S. student response with unjustified symbol manipulation in solving the Number Theory Problem.

was coded as a symbolic representation if a student mainly used mathematical expressions to explain how he or she found the answer. The response was coded as a visual representation if a student mainly used a picture or drawing to explain how he or she found the answer.

Figure 15 shows the percentage distributions of the U.S. and Chinese students who used various modes of representation. The frequencies were significantly different (x^2 (1, N = 558) = 105.97, p < .001). In particular, a larger percentage of Chinese than U.S. students used symbolic representations ($z = 9.84$, $p < .01$); whereas a larger percentage of U.S. than Chinese students used verbal representations ($z = 6.47$, $p < .01$) and visual representations ($z = 2.22$, $p < .01$).

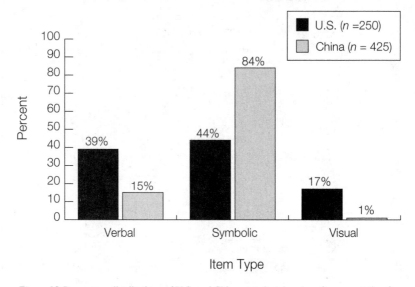

Figure 15. Percentage distributions of U.S. and Chinese students' modes of representation for solving the Number Theory Problem.

In summary, as in the Estimation Problem, both groups of students encountered great difficulty on the Number Theory Problem. In fact, the lowest mean scores for both groups occurred on this problem, and both the U.S. and Chinese students had similar performance levels on this problem (cf. Table 4). For students who had correct answers, U.S. students tended to provide a correct answer of 13, and Chinese students tended to provide correct answers other than 13. A larger percentage of U.S. than Chinese students used trial-and-error strategies, whereas a larger percentage of Chinese than U.S. students used common-multiples strategies. With respect to mathematical errors, a considerable number of U.S. and Chinese students manipulated numbers unreasonably to get an answer. A larger percentage of U.S. than Chinese students made errors that involved failing to satisfy the conditions of the problem. To represent their solution processes, U.S. students used verbal and visual representations more frequently than Chinese students, and Chinese students used symbolic representations more frequently than U.S. students.

Qualitative Results for the Pattern Problem

The fourth open-ended problem analyzed was the Pattern Problem (OE-5), where four initial figures of a pattern were given and students were asked to draw the fifth and seventh figures and then to describe how they knew what the seventh figure should be. Each student response to this problem was coded with respect to five aspects: (1) drawing figures, (2) solution strategy, (3) drawing errors, (4) quality of explanation, and (5) mode of explanation. The results are reported in five separate sections.

Drawing figures. Students were asked to draw the fifth and seventh figures for the pattern in this problem. A significantly larger percentage of Chinese (83%) than U.S. (73%) students drew both the fifth and seventh figures correctly ($z = 3.01$, $p < .01$). The vast majority of U.S. (87%) and Chinese (90%) students drew either the fifth or seventh figure correctly. About 10% of U.S. and 5% of Chinese students drew the fifth figure correctly, but not the seventh figure. Few U.S. and Chinese students drew only the seventh figure correctly.

Solution strategies. There were seven different solution strategies used by at least one U.S. or Chinese student in solving this problem.

Strategy 1
  Students focused on the number of dots in the three rows of each figure as a triplet and induced the rule of the triplets (Fig. 1: [1, 2, 3]; Fig. 2: [2, 3, 4]; Fig. 3: [3, 4, 5]; etc.).

Strategy 2
  Students looked at the dots in each row across the figures as a sequence and found the rule of each sequence (Row 1: {1, 2, 3, 4, ...}; Row 2: {2, 3, 4, 5, ...}; Row 3: {3, 4, 5, 6, ...}).

Strategy 3
  Students looked at the figures diagonally and realized that each successive figure had one more diagonal of 3 dots. By adding one more diagonal to the fourth figure, they would get the fifth figure in the pattern.

Strategy 4
  Students realized that, from figure to figure, each row has one more dot than the corresponding row in the previous figure. From the fourth figure, they added one dot to each row to get the fifth figure.

Strategy 5
  Students removed the first row of the previous figure and then added to it a new bottom row that had one more dot than the previous bottom row. This yielded the next figure.

Strategy 6
  Students found that the number of dots in the first row of a figure was equal to the number of each figure. The number of dots on the second row was one more than the first row, and the number of dots on the third row was one more than the second row.

Strategy 7
  Students focused on the total number of dots in each figure to describe how to get the next figure. In this case, they did not mention the shape of each figure explicitly and may actually have ignored the shapes of the figures.

A larger percentage of U.S. (83%) than Chinese (74%) students had apparent strategies ($z = 2.66$, $p < .01$). As shown in Table 13, the percentage distributions of the use of these strategies are quite similar. For those who had apparent strategies, about 40% of the U.S. and Chinese students used strategy 4, which was the most frequently used strategy for both U.S. and Chinese students. The second most frequently used strategy was strategy 7, which was employed by about one-fourth of both groups.

Table 13
Percentage Distributions of U.S. and Chinese Students with Apparent Strategies for the Pattern Problem

	Percentage of students	
Strategy	U.S. ($n = 208$)	China ($n = 316$)
Strategy 1	3	9
Strategy 2	7	8
Strategy 3	12	7
Strategy 4	41	37
Strategy 5	0[a]	1
Strategy 6	11	15
Strategy 7	25	24
Total	99[b]	101[c]

[a]Percentage is zero because of rounding.
[b]Percentage does not add to 100 because rounding.
[c]Percentage does not add to 100 because of rounding.

Drawing errors. Overall, 68 U.S. and 65 Chinese students drew either the fifth or seventh figure incorrectly, and these are the students included in the error analysis reported in this section. The majority of the drawing errors seemed to result from student difficulties in coordinating the two dimensions of the problem: the numbers of dots and the shapes of the figures. About 70% of U.S. (48 of 68) and 50% of Chinese (35 of 65) students' incorrectly drawn figures correctly showed one of these two dimensions. For example, about 20% of both groups correctly drew all 18 or 24 dots for the fifth or seventh figure respectively but did not maintain the shape of the figures. A few students drew the correct shape but did not maintain the correct number of dots. Another 50% of U.S. and 30% of Chinese students' drawing errors showed nearly correct attempts to coordinate the two dimensions; for example, students drew either 23 or 25 dots, arranged in the appropriate shape for the seventh figure. These may have been careless errors. About 30% of the U.S. students and 50% of the Chinese students' incorrect drawings showed serious errors in both the shape and the number of dots; for example, 60 dots might be drawn randomly.

Quality of explanation. The quality of a student's explanation involves the correctness and clarity of the written communication. In particular, the quality of students' written explanations were classified into one of the following five categories:

1. Complete and correct explanation: A student's explanation clearly indicated the processes used to get the seventh figure.

2. Partially correct: A student's explanation was incomplete, but correct. Only part of the solution process was described.

3. Procedural explanation: A student's explanation was procedural and lacked connection to the specific details of the solution process.

4. Description of the seventh figure: A student's explanation contained only what the seventh figure looks like, rather than how the seventh figure was obtained.

5. Incomprehensible description: A student's explanation made no sense.

Table 14 shows the percentage distributions of the U.S. and Chinese students who provided various types of explanations. The majority provided complete and correct, or partially correct, explanations. A slightly larger percentage of U.S. than Chinese students provided complete and correct explanations. If the percentages of the students with completely correct and partially correct explanations are combined, a significantly larger percentage of U.S. than Chinese students had completely correct or partially correct explanations ($z = 2.42, p < .01$).

Table 14
Percentage Distributions of U.S. and Chinese Students' Explanations for the Pattern Problem

Explanation types	Percentage of students	
	U.S. ($n = 250$)	China ($n = 425$)
Complete and correct	48	42
Partial correct	36	34
Procedural	9	12
Description of the seventh figure	3	0[a]
Incomprehensible explanation	4	5
No explanation	0[a]	5
Did not respond	0	1
Total	100	99[b]

[a]Percentage is zero because of rounding.
[b]Percentage does not add to 100 because of rounding.

About 10% of both the U.S. and Chinese students provided procedural explanations. A typical example is the following: "I looked at the four figures and found a rule, then I followed the rule to figure out the seventh figure." It is interesting to note that 5% of the Chinese students drew either the fifth or seventh figure (most of them were correct) but did not provide any explanations. Only one U.S. student did so. Only a few U.S. and Chinese students described what the seventh figure looks like rather than explaining how they got the seventh figure.

Mode of explanation. All but one U.S. student provided an explanation of how they knew what the seventh figure would be like on the basis of the pattern given. About 93% of the Chinese students provided explanations of how they found the seventh figure. For those who provided explanations, the vast majority provided verbal explanations (95%, or 236 of 249 U.S. students; and 99%, or 392 of 397 Chinese students). The remaining U.S. students used a combination of verbal and visual representations.

In summary, the quantitative analysis showed that among the seven open-ended problems, the U.S. and Chinese students performed the best on the Pattern Problem, with the U.S. students performing as well as the Chinese students (cf. Table 4). Although a larger percentage of Chinese than U.S. students drew both the fifth and seventh figures correctly, a larger percentage of U.S. than Chinese students had apparent strategies and provided complete or partial explanations of how they found the seventh figure. Both groups committed similar types of drawing errors, with the most frequent errors resulting from student difficulties in coordinating the two dimensions of the pattern (i.e., the numbers of dots and the shapes of the figures). Over 95% of both the U.S. and Chinese students used only verbal representations in their explanations of solutions.

Qualitative Results for the Average Problem

The last open-ended problem analyzed was the Average Problem. Each student's response to this problem was coded with respect to (a) numerical answer, (b) mathematical errors, (c) solution strategy, and (d) mode of representation.

Numerical answer. The numerical answer was what the student provided on the answer space for the task, and it was judged correct or incorrect. In a few cases, a student's explanation clearly supported the correct answer of 10, but the number recorded on the answer space was not 10. These answers were coded as correct. Three-fourths of the Chinese students had the correct answer of 10, which is significantly larger than for the U.S. students (one-half) ($z = 6.82, p < .001$).

Mathematical errors. Students who did not give the correct answer of 10 were subject to analyses of error types. Overall, 126 (50%) U.S. students and 100 (24%) Chinese students were included in the error analysis. Obviously, a larger percentage of U.S. than Chinese students made mathematical errors in solving this problem. The six different types of errors made in solving this problem are described below:

1. Minor calculation error: The student used the correct solution process for the problem, but made a minor calculation error.

2. Violation of "stopping rule": The student used a trial-and-error strategy but stopped trying when (a) the quotient was not 7; (b) the remainder was not zero; or (c) the quotient was not 7 and the remainder was not zero.

3. Incorrect use of average concept or formula: The student tried to apply the average concept or formula to solve the problem, but the application was incorrect. Below are a few examples.

   Example 1: The student added the numbers of hats sold in week 1 (9 hats), week 2 (3 hats), week 3 (6 hats), and the average (7 hats), then divided the sum by 4; the student then gave the whole number quotient (6 hats) as the answer.

   Example 2: The student added the numbers of hats sold in week 1 (9 hats), week 2 (3 hats), and week 3 (6 hats), divided the sum by 3, and then gave the quotient (6 hats) as the answer.

Example 3: The student added the numbers of hats sold in week 1 (9 hats), week 2 (3 hats), and week 3 (6 hats), then divided the sum by 3, and got 6. However, the average was 7. Therefore the student added 3 to the sum of the numbers of hats sold in week 1 (9 hats), week 2 (3 hats), and week 3 (6 hats), then divided it by 3, got 7, and then gave the answer 3.

Example 4: The student added the numbers of hats sold in week 1 (9 hats), week 2 (3 hats), and week 3 (6 hats), then divided the sum by 3, and got 6; 6 + 1 = 7, so the answer is 1.

4. Unjustified symbol manipulation: The student just picked some numbers from the task and worked with them in ways irrelevant to the problem context (for example, added them together).

5. The student gave the total number of hats sold in four weeks (28 hats) as the answer.

6. Errors cannot be identified: A student's work or explanation was so unclear or incomplete that the error type could not be identified.

Table 15 shows the number of U.S. and Chinese students who made mathematical errors. Similar numbers of U.S. and Chinese students made minor calculation errors and a considerable number of U.S. (30) and Chinese (22) students manipulated numbers in an unreasonable manner. Thirteen U.S. students violated the "stopping rule" when they used the trial-and-error strategy to solve the problem, but no Chinese student made this type of error. Forty-three (43) U.S. and 27 Chinese students incorrectly used the average formula. This represents 34% of the U.S. students who made mathematical errors and 27% of the Chinese students. These percentages are not significantly different. Consider the ratio of the number of students who incorrectly used the average formula to the total number of students who participated in the study; a larger percentage of U.S. (17%, or 43 of 250) than Chinese (6%, or 27 of 425) students used the average concept incorrectly ($z = 4.46$, $p < .001$).

Table 15
Frequency of U.S. and Chinese Students' Mathematical Errors When Solving the Average Problem

	Number of students	
Error types	U.S.	China
Minor calculation error	12	15
Violation of "stopping rule"	13	0
Incorrect use of average concept	43	27
Unjustified symbol manipulation	30	22
The total number of hats sold (28) as the answer	1	6
Errors cannot be identified	27	30
Total	126	100

Solution strategy. Only the responses of students who provided explanations were examined with regard to their solution strategies. This included 244 (98%)

of the U.S. students and 402 (96%) of the Chinese students. Three solution strategies used by at least one U.S. or Chinese student are described below.

Strategy 1 (Leveling)—The student solved the problem by adding or subtracting unit hats in order to "even off" the number of hats sold in 4 weeks at the same height, that of the average number of hats sold (7). Generally, students viewed the average number of hats sold (7) as a leveling basis to "line up" the numbers of hats sold in weeks 1, 2, and 3. Because 9 hats were sold in week 1, it has two extra hats. Because 3 hats were sold in week 2, an additional 4 hats are needed in order to line up the average. Because 6 hats were sold in week 3, 1 additional hat is needed to line up the average. In order to line up the average number of hats sold over 4 weeks, 10 hats should be sold in week 4.

Strategy 2 (Average formula)—The student used the average formula to solve the problem arithmetically (e.g., $7 \times 4 - (9 + 3 + 6) = 10$ or algebraically (e.g., $(9 + 3 + 6 + x) = 7 \times 4$, then solve for x).

Strategy 3 (Trial and error)—The student first chose a number for week 4 and then checked if the average of the numbers of hats sold for the 4 weeks was 7. If the average was not 7, then they chose another number for week 4 and checked again, until the average was 7.

Figure 16 shows examples for two of the solution strategies. In total, 73% of the U.S. students and 80% of the Chinese students used one of the strategies above. The percentage of Chinese students who had apparent strategies is significantly larger than that of U.S. students ($z = 2.23$, $p < .01$). Table 16 shows the percentage distributions of U.S. and Chinese students who used various solution strategies. For both U.S. and Chinese students, the most frequently used strategy was the average formula (Strategy 2). However, for those who had apparent solution strategies, the distributions between U.S. and Chinese students were significantly different (x^2 (2, $N = 523) = 65.25$, $p < .0001$). Chinese students tended to use the average formula to solve the problem more frequently than U.S. students, and U.S. students tended to use the trial-and-error strategy more frequently than Chinese students. Only a few in either group used the leveling strategy.

Table 16
Percentage Distributions of U.S. and Chinese Students Who Had Apparent Strategies for Solving the Average Problem

	Percentage of students	
Strategy	U.S. ($n = 182$)	China ($n = 341$)
Leveling strategy	4	1
Using average formula	65	92
Trial and error	31	6
Total	100	99[a]

[a]Percentage does not add up to 100 because of rounding.

It is important to note that not all students who had one of the apparent strategies used that strategy correctly. In fact, of the 182 U.S. students who had apparent

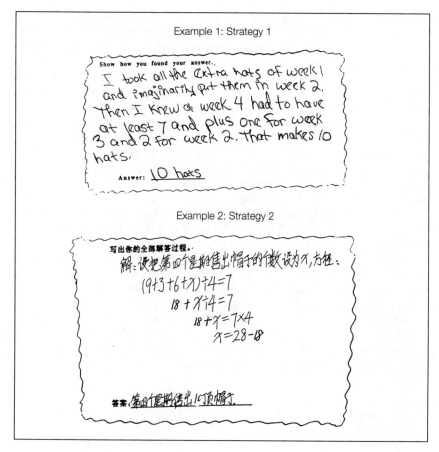

Figure 16. Examples of student solution strategies in solving the Average Problem.

strategies, 137 (75%) used the solution strategies correctly. Of the 341 Chinese students who had apparent strategies, 314 (92%) used the strategies correctly. The percentage of U.S. students who used the strategies correctly was lower than that for Chinese students ($z = 6.12$, $p < .001$). The majority of U.S. and all Chinese students who used the strategies incorrectly did so through the incorrect use of the average formula (strategy 2).

Mode of representation. Only student responses that contained explanations were examined for the mode of representation used. This included 244 (98%) U.S. students and 402 (96%) Chinese students. The mode of representation was examined in a way similar to that used in other tasks. The following categories were used to classify the mode of representation: verbal only, visual only, symbolic only, and any combination of these three. For this specific task, the symbolic representations were further classified into arithmetic symbolic and algebraic symbolic representations.

Table 17 shows the percentage distributions of U.S. and Chinese students' modes of representation when solving this problem. A larger percentage of U.S. than Chinese students used verbal-only representations ($z = 3.47, p < .001$) and verbal-related representations, including verbal only, combination of verbal and arithmetic symbolic, combination of verbal and algebraic symbolic, and combination of verbal and visual representations ($z = 5.98, p < .001$). However, a larger percentage of Chinese than U.S. students used symbolic-related representations including arithmetic symbolic only, algebraic symbolic only, combination of verbal and arithmetic symbolic, combination of verbal and algebraic symbolic, and combination of visual and arithmetic symbolic ($z = 6.02, p < .001$). If the arithmetic and algebraic symbolic representations were examined separately, the percentages of U.S. and Chinese students using arithmetic symbolic representations were very close (66% for U.S. students and 62% for Chinese students). But the percentage of Chinese students (26%) who used algebraic symbolic representations was much higher than that of U.S. students (2%) ($z = 7.84, p < .0001$). Therefore, the significant difference between the U.S. and Chinese students' use of symbolic representations was due to a higher percentage of Chinese students using algebraic symbolic representation. It is interesting to note that for both samples, the students who used algebraic symbolic representations performed significantly better than other students on computation tasks, simple problem-solving tasks, and on the remaining six open-ended problems.

About 13% of the U.S. students used visual-related representations, including visual only, a combination of verbal and visual, and a combination of visual and arithmetic symbolic representations in their solution processes. However, no Chinese students used this type of representation.

Table 17
Percentage Distributions of U.S. and Chinese Students' Various Modes of Representation for the Average Problem

Modes of representation	Percentage of students	
	U.S. ($n = 244$)	China ($n = 402$)
Verbal only	23	12
Visual only	2	0
Arithmetic symbolic only	45	52
Algebraic symbolic only	2	23
Combination of verbal and arithmetic symbolic	18	10
Combination of verbal and algebraic symbolic	0[a]	3
Combination of verbal and visual	7	0
Combination of visual and arithmetic symbolic	4	0
Total	101[b]	100

[a]Percentage is zero because of rounding.
[b]Percentage does not add to 100 because of rounding.

COMPARISONS OF STUDENTS' RELATIVE PERFORMANCE ON THE THREE TYPES OF ITEMS

In the previous two sections, U.S. and Chinese students' performance on the three types of items were compared. This section reports the results of comparisons of U.S. and Chinese students' relative performance on the three types of items. This section starts with an examination of the relatedness of their performance on the three types of items and ends with comparisons of their performance on the component questions and the open-ended problems when computation performance is matched.

Relatedness of Students' Performance on the Three Types of Items

Table 18 shows the correlation coefficients of U.S. and Chinese students' performance on the three types of items. For both U.S. and Chinese students, the correlation coefficients of their performance on any two types of items were moderately high. This suggested that there was a fairly strong relationship between students' performances on any two types of items. The correlation coefficients of the computation tasks with the component questions, and component questions with the open-ended problems, were of similar magnitude for U.S. and Chinese samples.

Table 18
Correlation Coefficients of Student Performance on Any Two Types of Items by Nation

	U.S. ($n = 250$)		China ($n = 425$)	
	Component	Open-ended	Component	Open-ended
Computation	.53	.61	.51	.49
Component		.62		.66

The evident difference between U.S. and Chinese student samples with respect to the correlation coefficients is in the correlation coefficient between the computation tasks and the open-ended problems. The correlation coefficient between computation tasks and open-ended problems for U.S. students is .61. This is significantly larger than the correlation coefficient for Chinese students (.49) ($z = 2.16$, $p < .01$). Thus, the relationship between U.S. students' performance on the computation tasks and the open-ended problems was stronger than that for Chinese students. However, the difference might be due to the smaller variation in Chinese students' performance on the computation tasks when compared with U.S. students.

For both the U.S. and Chinese students, several regression analyses were conducted to further examine the relationships between performance on the three types of items. Regression analyses (computation as a predictor and component questions as dependent variable) showed that U.S. and Chinese students' computation performance was a good predictor for performance on the component questions ($p < .001$). The amount of variance in performance on component questions accounted for by performance on computation tasks was very similar for both groups. Although regression analyses (computation as a predictor and open-ended problems as dependent variable) showed that U.S. and Chinese students' computation performance was also a good predictor of performance on the open-ended problems ($p < .0001$), the amount of variance on open-ended problems that was accounted for by Chinese students' performance on computation was different from that for U.S. students. Only 24% of the variance for Chinese students' performance on open-ended problems was due to variance of their computation performance, as compared to 37% of the variation for U.S. students' performance. This difference might also be due to the smaller variation of Chinese students' performance on the computation tasks.

Matched Comparisons

One of the research questions involved examining how U.S. and Chinese students performed on the component questions and open-ended problems when their computation performance was adjusted or matched. Blocking analysis was used to answer this research question where student performance on the computation tasks was used as a blocking factor, with similar numbers of U.S. and Chinese students and with the similar levels of the performance on computation selected in each block. Thus, the blocking matched U.S. and Chinese students on computation performance, and then compared their performance on the component questions and the open-ended problems.

In the blocking analysis, seven blocks were selected. The first block consisted of students whose score on the computation tasks was less than or equal to 14, the second block consisted of students whose score was 15, the third block consisted of students whose score was 16, the fourth block was 17, the fifth block was 18, the sixth block was 19, and the seventh block consisted of students whose score on the computation tasks was 20. For each block, if there were more than 10 students, 10 students were randomly selected. Table 19 shows the score levels on computation tasks and the number of U.S. and Chinese students in each block. Obviously, the selection processes of U.S. and Chinese students in each block led to the matched computation performance between U.S. and Chinese students in the blocking analysis.

Table 19
Computation Performance Levels and the Number of Students in Each Block

Blocks	Computation scores[a]	Number of students per block	
		U.S.	China
Block 1	< 14	10	10
Block 2	15	10	10
Block 3	16	10	10
Block 4	17	10	10
Block 5	18	10	10
Block 6	19	10	10
Block 7	20	9	10

[a]Possible computation score is 20.

Table 20 shows the mean scores of U.S. and Chinese students on the component questions and open-ended problems in each block. In almost every block, U.S. students had higher means than Chinese students on the component questions and open-ended problems.

Table 20
Mean Scores of U.S. and Chinese Students' Performance on Component Questions and Open-Ended Problems in Each Block When Computation Proficiency Is Held Constant

Blocks	Mean scores Component		Mean scores Open-ended	
	U.S.	China	U.S.	China
Block 1	11.2	9.5	17.3	10.8
Block 2	14.1	10.0	20.6	11.6
Block 3	14.2	10.2	22.0	7.4
Block 4	13.3	13.1	19.8	11.3
Block 5	13.2	13.2	21.0	15.3
Block 6	15.4	16.0	24.4	17.7
Block 7	17.0	17.0	24.6	22.6

Two separate ANOVAs were conducted to examine the differences between U.S. and Chinese students' performance on the component questions and the open-ended problems, when their computation performance was matched through the blocking procedure. The first ANOVA showed that U.S. students performed significantly better than Chinese students on the component questions when their computation performance was matched (F (7, 131) = 10.54, $p < .0001$). Table 21 shows the summary of the ANOVA for the component questions.

Table 21
Summary of the First ANOVA for the Component Questions

Source	DF	Sum of squares	Mean squares	F-value	Pr > F
Model	7	666.48	95.24	10.54	0.0001
Error	131	1183.29	9.04		

As with the previous ANOVA, the second ANOVA showed that U.S. students performed significantly better than Chinese students on the open-ended problems when their computation performance was matched (F (7, 131) = 19.31, $p < .0001$). Table 22 shows the summary of the ANOVA for the open-ended problems. Therefore, for those who had the same levels of computation performance, U.S. students had higher mean scores than Chinese students on both the component questions and the open-ended problems.

Table 22
Summary of the Second ANOVA for the Open-Ended Problems

Source	DF	Sum of squares	Mean squares	F-value	Pr > F
Model	7	3470.71	495.82	19.31	0.0001
Error	131	3363.74	25.68		

5. DISCUSSION

This study examined comparable groups of U.S. and Chinese students' mathematical performance on tasks involving computation, simple problem solving, and complex problem solving. A set of 20 multiple-choice arithmetic computation items was used to assess computation skills. Another set of 18 multiple-choice word-problem-solving component questions was used to assess simple problem-solving skills. In addition, a set of 7 open-ended problems was used to assess complex problem-solving skills.

The best effort has been made to have roughly equivalent U.S. and Chinese samples in this study, but the samples were not drawn to be nationally representative. Only 425 Chinese students from both common and key schools and 250 U.S. students from both private and public schools were sampled. Nor was the sampling of subjects based on random stratified sampling techniques at the research sites. Rather, all students in each volunteer class were studied. In this sense, a sample of convenience with safeguards was used in the research, and caution is needed in explaining the relative performance differences, discussing the possible implications of the findings, and generalizing the findings of the study. Despite the sampling limitations, the findings from a study that includes the in-depth cognitive analysis of performance should still be informative and useful in educational research and policy development (c.f., Bradburn & Gilford, 1990).

PERFORMANCE DIFFERENCE PATTERNS

Results of this study suggest that Chinese students performed significantly better than U.S. students on computation and simple problem solving but not on complex problem solving. In particular, Chinese students not only had higher mean scores than the U.S. students on computation tasks and component questions but also a higher percentage of perfect scorers. In contrast, on the open-ended problems, the U.S. and Chinese students had almost identical mean scores and almost identical percentages of the perfect scorers.

For computation and simple problem solving, the patterns of performance differences between the U.S. and Chinese students in this study were similar to those reported in previous cross-national studies that involved U.S. and Chinese students. In the previous cross-national studies (e.g., Lapointe et al., 1992; Stevenson et al., 1990), Chinese students far outperformed their U.S. counterparts. Like the students in those studies, the Chinese students in this study were much more successful than U.S. students in solving computation tasks and component questions. However, what was left unexamined, and hence unreported, in the previous cross-national studies is that the performance differences between U.S. and Chinese students appear to be related to the types of tasks and performances being examined. In this study, there was essentially no difference between U.S. and Chinese student performance on the open-ended problems, a finding rarely reported in the previous cross-national studies in mathematics.

Results of this study provide strong and direct evidence for supporting the hypothesis, suggested by Cai and Silver (1995), that the magnitude of the performance differences between U.S. and Chinese students on tasks measuring mathematical understanding or mathematical applications might be smaller than the differences that have been shown to exist for tasks measuring procedural knowledge or computational skills. This hypothesis was supported by the relatively larger performance difference between U.S. and Chinese students on the computation tasks, the smaller difference on the component questions, and the absence of a significant difference on the open-ended problems. In addition, further evidence was provided in the qualitative analyses of one of the open-ended problem: the Division Problem (OE-1). In solving the Division Problem, one not only needs to correctly apply and execute division computation (computation phase) but also to interpret correctly the computational results with respect to the given situation (sense-making phase). As suggested from prior research (Cai & Silver, 1994, 1995), Chinese students did significantly better than U.S. students on the computation phase. However, on the sense-making phase, Chinese students did not do better than U.S. students.

Hatano (1988) indicated that having expertise in routine applications does not imply expertise in complex and novel problem solving. Routine applications require mainly procedural knowledge, whereas complex and novel problem solving requires mainly conceptual knowledge (Hiebert, 1986). This study provided empirical data in support of Hatano's assertion. In particular, this assertion was supported by Chinese students' varied performance on the three types of tasks, with the highest performance on computation and lowest performance on complex problem solving.

There are several plausible interpretations for the finding that Chinese students outperformed U.S. students on the computation tasks and component questions but not on the open-ended problems. The findings might be due to different emphases in U.S. and Chinese school curricula and classroom instruction. Chinese students did best on the computation tasks, second best on the component questions, and lowest on the open-ended problems, but U.S. students had a similar performance level across all three types of tasks. The curriculum and classroom instruction in China might emphasize computation more heavily than mathematical understanding and application; whereas the U.S. curriculum and classroom instruction might emphasize computation and mathematical understanding equally, though each with somewhat less intensity than Chinese instruction related to computation and simple problem solving. In fact, although most of the American and Chinese teachers indicated that their students should have had knowledge to answer the majority of the tasks, Chinese teachers felt that their students were more familiar with the task format for the computation tasks than for the other two sets of tasks. To examine this hypothesis regarding the relatedness of differential performance patterns to curriculum and instruction, future studies are needed to document how procedural and conceptual knowledge are taught in American and Chinese schools.

As seen in Chapter 2, a number of factors might be taken into account to interpret the performance differences between students in China and the United States. For example, the performance difference patterns might be related to the fact that Chinese students were more interested in computation than complex problem solving, whereas U.S. students were equally interested in computation, simple problem solving, and complex problem solving. It is also possible that the pattern is related to teachers' beliefs about excellence of mathematics. Chinese teachers may believe that the abstract nature of mathematics is an elegant part of it, so they would like their students to practice abstract mathematics. In fact, Chinese students not only performed better than U.S. students on computations, but they also used abstract symbolic representations more frequently than American students in solving some of the complex problems.

A second possible interpretation for the finding that Chinese students did not outperform U.S. students on the open-ended problems might be related to the operations embedded in these open-ended problems being simple computations with whole numbers. Results from the item analysis of the computation tasks suggested that the U.S. students' poor performance on the computation tasks was due largely to their difficulties in solving tasks with fractions and some decimals. The difference between U.S. and Chinese students on nonfraction computations was smaller than that on the fraction computations, and the computations embedded in five of the open-ended problems involved only simple operations with whole numbers. The other two open-ended problems (Pattern Problem [OE-5] and Estimation Problem [OE-2]) did not necessarily require computations. Therefore, the combination of relatively easy numbers and operations in the five open-ended problems might have allowed the U.S. students to perform as well as the Chinese students. Interestingly, the mean differences between U.S. and Chinese students were the smallest on the Estimation Problem and the Pattern Problem. The implication is that they actually performed equally well on the open-ended problems that required essentially no computation.

A third possible explanation of the relatively small performance difference on the open-ended problems might be the fact that many Chinese students did not respond to the open-ended problems. Not only was there a larger percentage of Chinese students than U.S. students who did not respond to at least one open-ended problem, but there was also a larger percentage of Chinese students than U.S. students who did not respond to each of the open-ended problems. Once students with nonresponses were excluded from the analysis, Chinese students had significant (statistically) larger mean scores than U.S. students on the open-ended problems. Thus, of those students who attempted the open-ended problems, the performance of Chinese students was better than U.S. students.

NONATTEMPTED RESPONSES

The high rate of Chinese students' nonattempted responses to the open-ended problems was an unexpected finding in this study, because no previous cross-national study reported that Asian students in general, and Chinese students in particular, had lower response rates than U.S. students. In fact, a number of studies (e.g., Berry et al., 1992; Lynch & Hanson, 1992) have suggested that Asian students, including Chinese students, are more persistent than U.S. students. Mathematics teachers in Asian countries believe that the more a student struggles, the more likely it is that the student will learn (Stigler & Perry, 1988). Given previous reports, it is surprising that Chinese students had such low response rates.

Why did the Chinese students have a lower response rate than U.S. students? Is it possible that the Chinese students had less time than the U.S. students to complete these open-ended problems? The Chinese version of the administration procedures was literally translated from the English version, and an English back-translation was done to ensure translation equivalence. A follow-up inquiry found that Chinese students did have exactly 40 minutes (reading administration procedures was not included in this 40 minutes) to complete these problems, as did the U.S. students. Therefore, it is unlikely that the Chinese students had less time than the U.S. students. Further evidence that time was not a factor stems from the fact that the Chinese students did not show a higher response rate for the problems presented in the beginning than for those at the end of the task booklet, assuming that they would probably work through the problems in order in the booklet.

An alternative explanation of Chinese students' low response rate is the degree of difficulty and novelty of the problems. The open-ended problems are cognitively complex problems, although the numbers and operations embedded in the problems are simple. The low response rate might be due to the complexity of the open-ended problems for the Chinese students. Chinese students are rarely encouraged to make a blind guess. Thus, they might feel more comfortable leaving a difficult problem blank than having to guess. In contrast, U.S. students demonstrated a greater willingness to take risks. Some evidence to support this argument comes from a multiple-choice computation task involving negative numbers, a topic that neither sample had been taught. A fairly large proportion of the Chinese students were not willing to take the risk of choosing an answer for the computation task. Although U.S. students were not taught addition with a negative number, only a small proportion of them refrained from choosing an answer to a question that involved such knowledge.

Moreover, for the open-ended problems, although none, or a small proportion, of the U.S. students did not respond to them, U.S. students had a higher percentage of "0" scores than Chinese students for almost every problem, even for those problems on which U.S. students performed better than or as well as Chinese students. In contrast, when Chinese students had a low mean score for a problem, they also tended to have a low response rate. In general, if a student attempted a problem but the response showed no understanding of the problem, the response would be scored as "0". Thus, for those U.S. and Chinese students who did not have a clear sense of how to attack a problem, Chinese students may have more commonly skipped the problem, whereas U.S. students tended to write down some possibly related ideas. This may account for the fact that U.S.

students tended to have, overall, a higher quantity of attempted responses than Chinese students but that, for those who attempted a response, Chinese students tended to have higher quality responses than U.S. students.

RELATIVE PERFORMANCE DIFFERENCE: EXPLANATION AND IMPLICATION

This study extended the work of Mayer et al. (1991), who examined U.S. and Japanese fifth graders' mathematical performance by using computation and component questions. Mayer et al. (1991) reported that Japanese students outperformed their U.S. counterparts on both types of problems. But an additional analysis showed that when samples of U.S. and Japanese students were equated on the basis of their level of computational performance, U.S. students scored higher than comparable Japanese students on the component questions. This study extended that work by also including open-ended problems.

This study used a blocking method to match U.S. and Chinese students on computation performance. Despite the methodological differences between the study by Mayer et al. (1991) and this study in matching students on computation performance, the findings were quite similar. In particular, when subsets of U.S. and Chinese students were matched on their computation performance, U.S. students scored significantly higher than comparable Chinese students on measures of simple and complex problem solving.

The relative performance differences of U.S. and Chinese students on the three types of items can be explained in both statistical and educational terms. In the blocking analysis, U.S. and Chinese students in each block were matched on the basis of their computation performance. Recall that Chinese students had an very negatively skewed distribution and U.S. students had a relatively flat distribution on computation performance. Therefore, in the blocking analysis, only a small proportion of those Chinese students with high computation scores, but a large proportion of those Chinese students with low computation performance, were chosen and compared with their U.S. counterparts. Almost all the U.S. students with high computation performance and a small proportion of those with low computation performance were included in the analysis. Thus, a large proportion of U.S. students with low computation performance and a large proportion of Chinese students with high computation performance were excluded from the blocking analysis. Therefore, it is not surprising that the U.S. students performed better than the Chinese students on component questions and open-ended problems after matching their performance on computation tasks.

The relative differences between U.S. and Chinese students on the component questions and open-ended problems can also be interpreted from an educational perspective. Overall, Chinese students may receive more opportunity to learn basic mathematical skills, such as computation and simple word-problem solving, than do U.S. students. However, U.S. and Chinese students may have similar opportunity to learn complex problem-solving skills. Results from this study support the relative exposure hypothesis. U.S. students had very similar performance levels for all three types of problems, a performance level that was about the same as the Chinese on the complex problems, but lower on the other two types of tasks. This may suggest that the ratio of emphasis on computation to emphasis on simple word-problem solving and complex problem solving in Chinese schools may be higher than in U.S. schools.

Previous cross-national studies did not consistently show that students in the U.S. spend less time studying mathematics than students in Asian countries (Chen & Stevenson, 1990; Lapointe et al., 1992; Robitaille & Garden, 1989). In comparisons of U.S. and Chinese students, however, previous cross-national studies (Chen & Stevenson, 1990; Lapointe et al., 1992) consistently showed that Chinese students had not only more school days per year than U.S. students, but they also spent more time on studying mathematics in school and doing mathematics homework. In order to improve U.S. students' mathematical performance, it has been suggested that U.S. students' study time in mathematics be increased (Anderson & Postlethwaite, 1989; Walberg, 1988). Walberg indicated that "the amount of time devoted to schooling in the U.S. has increased substantially during the 20th century, but it is difficult to argue that the time allocated is yet sufficient given the increasing cognitive demands of the job market, poor achievement scores of U.S. students by international standards, and the average of 28 hours per week they spend watching television" (1988, p. 85). Thus, raising time allocations and engaging students for a greater fraction of allocated time is likely to help U.S. students' learning. The question is, How should the extra time be invested? There are at least three plausible ways to use this time: (a) solely on computation and simple problem solving; (b) solely on complex problem solving; and (c) distributing the time across all three types of activity—computation, simple problem solving, and complex problem solving.

The results of this study suggest that U.S. mathematics education does as good a job on computational skills as on simple problem solving and complex problem solving. The results also show that U.S. mathematics education does as well as Chinese mathematics education on the development of students' complex problem-solving skills. And when U.S. and Chinese students are matched on their computation performance, U.S. students actually perform better than Chinese students. One may argue that if U.S. students were exposed to basic mathematical skills such as computation as much as Chinese students, U.S. students' basic mathematical skills could be improved and approach a level equal to that of Chinese students. That is, if extra time were invested solely on computation and simple problem solving, U.S. students' performance on computation and simple problem solving might approach a level equal to their Chinese counterparts. It is possible that U.S. students' performance on computation and simple problem solving would approach a level equal to that of Chinese students but that their complex problem solving would not similarly improve. The results obtained for the Chinese sample in this study show that such a performance profile may be obtained. However, investing time in this way might be contrary to the thinking of leaders in the reform movement in U.S. mathematics education.

The NCTM *Standards* (NCTM, 1989, 1991) suggest that school mathematics curricula should deemphasize basic mathematical skills such as computation. NCTM (1989) also suggests decreasing the attention paid to complex paper-and-pencil computations, addition and subtraction without renaming, long division, paper-and-pencil fraction computation, and teaching computation out of context. At the same time, NCTM (1989, 1991) stresses the importance of the development of mathematical understanding. The deemphasis on computation and increased emphasis on mathematical understanding is aimed at developing students' mathematical reasoning and higher-level thinking skills. This suggests a second way to invest the extra time for mathematics—namely, to focus solely on complex problem solving. Would it be possible for students to develop their mathematical reasoning and higher-level thinking skills with no, or only poor, computation skills? It is reasonable to argue that basic mathematical skills are necessary for students to improve their higher-level thinking skills. For example, students' computation skills are closely related to the development of their number sense and operation sense (NCTM, 1989). Thus, it does not follow from NCTM recommendations that the extra time be invested solely in focusing on complex problem solving.

A third possible use of extra time for mathematics instruction would be to focus on the development of students' complex problem solving, but not solely on it. One of the main criticisms of existing U.S. mathematics programs is their overemphasis on procedural knowledge or symbols (e.g., Davis, 1986). The U.S. student performance pattern found in this study—similar levels of performance on complex problem solving, simple problem solving, and computation—might reflect the influence of current ideas of mathematics education reform. However, because almost all the U.S. students who participated in this study attended private schools, it might also reflect the kind of thoughtful mathematics education made available to elite groups in the U.S. but not generally available to all students (Silver, 1994). The use of more time to provide a balanced treatment of computation, simple problem solving, and complex problem solving in school mathematics would not only reflect the spirit of calls for mathematics education reform but would also have great potential to improve students' mathematical performance on all three types of tasks. Consequently, it might be conceivable that all students would perform at a much higher level on computation, simple problem solving, and complex problem solving.

COGNITIVE SIMILARITIES AND DIFFERENCES
IN COMPLEX PROBLEM SOLVING

Although U.S. students did not differ from Chinese students in their mean score for complex problem solving, the detailed cognitive analysis of student responses to the open-ended problems revealed many subtle differences and similarities in the thinking and reasoning between the two samples of students. These differences and similarities would not have been revealed if open-ended problems were not included in this study, or if the detailed cognitive analyses of student responses to open-ended problems were not conducted. The cognitive analyses provided not only detailed information about the similarities and differences between U.S. and Chinese students in their solving of the open-ended problems, but they also demonstrated the feasibility of the analysis to capture insights into students' thinking.

Several similarities and differences between U.S. and Chinese students were revealed. For example, in some cases, both samples of students used similar strategies to obtain their correct numerical answers, good estimates, or correct figures; in other cases, they did not. In solving the Number Theory Problem (OE-4), which allowed for multiple correct answers, U.S. and Chinese students tended to provide different correct answers. Although the percentages of U.S. and Chinese students who had correct answers were almost identical, further examination showed that the nature of their correct answers was different. In particular, a significantly larger percentage of U.S. students than Chinese students gave a correct answer of 13, but a larger percentage of Chinese students than U.S. students gave a correct answer of 25. The different frequencies of correct answers is due to the fact that U.S. and Chinese students used different strategies. The U.S. students used strategies that would yield a correct answer of 13 more frequently; Chinese students used strategies that would yield a correct answer of 25 more frequently. This finding suggests the importance of examining students' mathematical problem solving beyond correctness. In fact, if the Number Theory Problem had been presented using a multiple-choice format, only the correctness of answers could have been evaluated, and all the preceding information would have been missed.

There were similarities and differences between U.S. and Chinese students in their solution strategies, mathematical errors, and modes of representation. Almost every solution strategy used by U.S. students was also used by Chinese students, and vice versa. Moreover, the solution strategy used most frequently by U.S. students to solve five of the open-ended problems was also the strategy used most frequently by Chinese students. For example, in solving the Average Problem, the average formula was the most frequently used strategy by both U.S. and Chinese students, even though overall the Chinese students performed much better than the U.S. students. In solving the Division Problem, the vast majority of the U.S. and Chinese students used division computation, and in this case U.S. students outperformed Chinese students.

Although there were many similarities in the use of solution strategies, there were also some clear differences. The most notable difference is that U.S. students tended to use trial and error much more frequently than Chinese students, when it was an appropriate way to solve the problem. Second, U.S. students tended to use strategies related to visual representation more frequently than Chinese students. For example, in solving the Number Theory Problem, for those who showed apparent strategies, about 25% of the U.S. students found the common multiples of 2, 3, and 4 by drawing blocks, whereas only 1% of the Chinese students did. Also, Chinese students tended to use solution strategies correctly more often than U.S. students. This was evident in their attempts to solve the Average Problem. A considerable number of U.S. students used the average formula incorrectly. Moreover, many U.S. students did not carry the trial-and-error approach to its logical conclusion (i.e., many of them stopped further trials, although their previous check showed that the needed conditions had not yet been met). In solving the Average Problem, the incomplete use of the trial-and-error approach may be due to U.S. students' lack of conceptual understanding of the average concept, but it may also be due to their lack of knowledge about the approach itself.

A comparison of U.S. and Chinese students' errors in solving the open-ended problems was also quite revealing. Students from both samples had many errors in common. The most evident error type in common was that a considerable number of U.S. and Chinese students manipulated numbers from the problem and worked with them in ways irrelevant to the problem context. Many other errors resulted from minor calculation errors and a failure to meet problem conditions. Even though Chinese students outperformed U.S. students in the computation and component questions, item analysis showed that they had errors in common. However, Chinese students tended to make careless calculation errors more frequently than U.S. students, whereas U.S. students tended to make errors due to failing to satisfy problem conditions more frequently than Chinese students.

With respect to mode of representation, both U.S. and Chinese students tended to use verbal representations more often than symbolic and visual representations to explain or show how they found their solutions. In some cases, the frequent use of verbal representations might be related to the words used in the directions, such as "explain" and "describe." Students' understanding of the terms "explain" and "describe" may imply a need to use words. For example, for the Pattern Problem, students were asked to "describe how you knew what the seventh figure would look like." Over 95% of both the U.S. and the Chinese students provided verbal descriptions. And verbal representation was still one of the most frequently used modes in the tasks directing students to "show how you found your answer."

With respect to students' usage of different modes of representation, U.S. students tended to use visual or pictorial representations more frequently than Chinese students, and Chinese students used symbolic or notational representations more frequently than U.S. students. This difference was dramatically shown in the solutions to the Number Theory Problem and the Average Problem. In solving the Number Theory Problem, the vast majority of the Chinese students who attempted the problem used symbolic representation (i.e., mathematical expressions). Only 10% of the Chinese students used a verbal representation, and only a few students used a visual representation. However, nearly 20% of the U.S. students used a visual representation, with the rest of the students equally distributed between verbal and symbolic representations.

Similarly, in solving the Average Problem, over 10% of U.S. students' explanations involved visual representation, but no Chinese student used this type of representation. Chinese students tended to use symbolic representations more frequently than U.S. students, but quite a large proportion of U.S. students (64%) used some symbolic representations. Further examination of the symbolic representations used in student explanations of solutions suggested that a similar proportion of U.S. and Chinese students used arithmetic symbolic representation, but a larger proportion of Chinese than U.S. students used algebraic symbolic representation. Thus, Chinese students not only used symbolic representation more frequently than U.S. students, but Chinese students also used more abstract symbolic representation than U.S. students.

These findings are similar to those reported by Silver et al. (1995) in their examination of U.S. and Japanese students' solutions to a mathematical problem. They found that the U.S. students tended to use visual representations more often than the Japanese students, whereas the Japanese students expressed their thinking using symbolic representations more frequently than the U.S. students. Moreover, for those who used symbolic representations, U.S. students tended to use addition more frequently than multiplication, but Japanese students tended to use multiplication more often than addition. A similar finding can also be found in Ito-Hino (1994).

It is reasonable to assume that the differences between U.S. and Chinese students in their use of modes of representation are related to their classroom experiences. For example, U.S. teachers, more than Chinese teachers, may use manipulative and concrete examples or materials in classrooms in helping students understand mathematics concepts. Chinese schools are encouraged to use concrete examples and materials in mathematics classrooms, but their role is clearly to mediate an understanding of mathematical abstractions (Cao & Cai, 1989). The U.S. students' relatively less frequent use of the abstract mathematical expressions and more frequent use of the concrete visual representations may suggest that U.S. teachers fail to carry the use of concrete materials to the point where students reach the abstraction level. In fact, U.S. teachers tend to believe that young children need concrete experiences in order to understand mathematics, at times asserting that concrete experiences would automatically lead to understanding (Stigler & Perry, 1988). Examination of how U.S. and Chinese teachers use various modes of representation and how their instructional emphasis influences their students' uses of different modes of representation are needed in future cross-national investigations.

Because symbolic representation is more abstract than verbal or visual representations, it might be reasonable to assume that students who used symbolic representation performed better than those who used verbal and visual representations because of the abstract nature of mathematics. In general, students in both samples who used verbal representation did not perform as well as those who used symbolic and visual representations. In other words, students who used symbolic representation performed better than those who used other representations. For those U.S. and Chinese students who used symbolic representation, the more abstract the representation they used, the better they performed. These findings suggest that students may benefit from being asked to find general solutions to mathematical problems and to represent those solutions and their ways of thinking by using formal mathematical symbolism. Of course, such benefit can only be gained if mathematics teachers provide appropriate mathematical problems and encourage students to search for general solutions to the problems.

In this study, because so few Chinese students used visual representations, no analyses were conducted to compare the performance of students who used visual representations with Chinese students who used other representations. In comparisons between U.S. students who used visual representation and those who used other representations, the results based on the Number Theory Problem and the Average Problem were somewhat different. In the Number Theory Problem, U.S. students who used visual representations had higher means on the computation tasks, component questions, and open-ended problems than those who used symbolic representations, although the differences were not statistically significant. For the Average Problem, however, U.S. students who used visual representation had lower means on the computation tasks, component questions, and open-ended problems than those who used arithmetic symbolic and algebraic symbolic representations, and most of the differences were statistically significant. Given these differential results, additional investigations are needed to examine how students who use visual representations perform when compared to the students who use other representations.

INTEGRATING COGNITIVE ANALYSIS AND PERFORMANCE ASSESSMENT

This study attempted to integrate cognitive psychology and performance assessment in a cross-national study. Importantly, the study used a variety of mathematical tasks to capture the thinking and reasoning of U.S. and Chinese students. The advantage of this approach is that different mathematical tasks allow measurement of different kinds of mathematical performance. Thus, a cross-national study using a wide array of mathematical tasks can reveal not only similarities and differences in global performance but also detailed information about these similarities and differences. In particular, this study revealed similarities and differences between U.S. and Chinese students' thinking and reasoning in contexts beyond computation and simple problem solving.

Another defining attribute of this study is that it analyzed students' performance not solely on the percentage of correct or incorrect responses but rather on detailed analyses of student strategies, representations, and errors. The findings concerning subtle cognitive similarities and differences between U.S. and Chinese students with respect to solution strategies, mathematical errors, and modes of representation were only possible because a detailed cognitive analysis of student responses to the open-ended problems was conducted. The range of tasks used and the methodology employed supported the discovery of the unique findings of cognitive similarities and differences between U.S. and Chinese students that have not previously been reported.

Moreover, this study focused on both similarities and differences between U.S. and Chinese students' mathematical performance. As mentioned in Chapter 2, most previous cross-national studies focused only on examining student performance differences followed by an examination of differences on cultural and educational factors. After identifying any such differences, researchers tried to use them to explain existing student performance differences. The focus here on similarities, as well as differences, not only led to a consideration of the possible factors contributing to the performance differences but also to an examination of why and how learning in different circumstances sometimes led to similar outcomes. Of course, the findings here are limited to the three types of mathematical tasks and the analysis scheme used. The analysis scheme, however, can serve as a prototype for further investigations involving other types of tasks.

The purpose of this study was to provide in-depth information about U.S. and Chinese students' thinking and reasoning. No direct background information was collected to explain the observed similarities and differences between the two groups. It would certainly be valuable to document classroom instruction in U.S. and Chinese schools and to examine how it is related to the observed differences in performance between U.S. and Chinese students. The success of integrating cognitive psychology and performance assessment into this cross-national study suggests the feasibility of integrating cognitive theories and methodology on teaching into cross-national studies of teachers' cognition and instruction.

Although some research has been done to document what actually happens in U.S. and Asian classrooms on the basis of informal and formal observations (e.g., Stigler et al., 1990), no cross-national studies have been conducted to describe how teachers' actions or teaching behaviors are correlated with students' success in thinking and reasoning (Brophy & Good, 1986). Furthermore, there are also no cross-national studies that analyze the complex cognitive processes of mathematics teaching in U.S. schools compared to that in Asian schools. A natural extension of this study is to document and analyze the complex cognitive processes of U.S. and Chinese mathematics teachers' classroom teaching and then to examine how their cognitive behaviors are correlated with their students' thinking and reasoning. Fortunately, some researchers (e.g., Leinhardt, 1993; Shulman, 1986) have developed effective research paradigms and methods for the study of teachers' cognition and instruction. These research paradigms and methods could be integrated into such comparative studies of teachers' cognition. Furthermore, the research project of classroom instruction in Japan and the U.S. carried out by Stigler (Stigler & Fernandez, 1993) may also provide insights for studying teachers' cognition.

The results of this study suggest both the complexity of examining mathematical performance differences and the feasibility of conducting such an examination in a multidimensional manner. It is hoped that this study not only contributes to our understanding of cross-national differences in mathematical performance but also suggests some productive directions for future cross-national studies. In particular, it is recommended that future cross-national performance comparisons use a wide array of mathematical tasks and an associated set of detailed cognitive analyses that capture the thinking and reasoning of students. The availability of new ideas and techniques from cognitive psychology and performance assessment herald a new era for cross-national studies of mathematical performance, an era that will deepen our understanding of student learning, problem solving, and reasoning.

REFERENCES

Aiken, L. R., Jr. (1979). Update on attitude and other affective variables in learning mathematics. *Review of Educational Research, 46*(2), 293–311.

Allwood, C. M. (1984). Error detection processes in statistical problem solving. *Cognitive Science, 8*, 413–437.

Anderson, J. R. (1985). *Cognitive psychology and its implication.* Hillsdale, NJ: Erlbaum.

Anderson, J. R. (1987). Skill acquisition: Compilation of weak-method problem solutions. *Psychological Review, 94*, 192–210.

Anderson, L. W., & Postlethwaite, T. N. (1989). What IEA studies say about teachers and teaching. In A. C. Purves (Ed.), *International comparison and educational reform* (pp. 51–72). Alexandria, VA: Association for Supervision and Curriculum Development.

Antonouris, G. (1988, September 30). Multicultural perspectives: Is math really "culturally neutral?" *The Times Educational Supplement.* p. 64.

Baranes, R., Perry, M., & Stigler, J. W. (1989). Activation of real-world knowledge in the solution of word problems. *Cognition and Instruction, 6*(4), 287–318.

Becker, J. P. (Ed.). (1992). *Report of U.S.-Japan cross-national research on students' problem solving behaviors.* Carbondale, IL: Southern Illinois University.

Becker, J. P., Silver, E. A., Kantowski, M. G., Travers, K. J., & Wilson, J. W. (1990). Some observations of mathematics teaching in Japanese elementary and junior high schools. *Arithmetic Teacher, 28*(2), 12–21.

Bejar, I. I., Embretson, S., & Mayer, R. E. (1987). *Cognitive psychology and the SAT: A review of some implications* (Rep. No. 87–28). Princeton, NJ: Educational Testing Service.

Berry, J. W., Poortinga, Y. H., Segall, M. H., & Dasen, P. R. (1992). *Cross-cultural psychology: Research and applications.* New York: Cambridge University Press.

Bjorklund, D. F. (Ed.). (1990). *Children's strategies: Contemporary views of cognitive development.* Hillsdale, NJ: Erlbaum.

Blinco, P. N. (1991). Task persistence in Japanese elementary schools. In E. R. Beauchamp (Ed.), *Windows on Japanese education* (pp. 51–75). New York: Greenwood Press.

Bracey, G. W. (1992). The second Bracey report on the condition of public education. *Phi Delta Kappan, 74*(2), 104–117.

Bracey, G. W. (1993). The third Bracey report on the condition of public education. *Phi Delta Kappan, 75*(2), 104–117.

Bradburn, M. B., & Gilford, D. M. (1990). *A framework and principles for international comparative studies in education.* Washington, DC: National Academic Press.

Brickman, W. W. (1988). History of comparative education. In T. N. Postlethwaite (Ed.), *Encyclopedia of comparative education and national systems of education* (pp. 3–9). Oxford: Pergamon.

Brislin, R. W. (1986). The wording and translation of research instruments. In W. J. Lonner & J. W. Berry (Eds.), *Field methods in cross-cultural psychology* (Vol. 2, pp. 389–400). Beverly Hills, CA: Sage.

Broadfoot, P., Murphy, R., & Torrance, H. (Eds.). (1990). *Changing educational assessment: International perspectives and trends.* New York: Routledge.

Brophy, J. E., & Good, T. L. (1986). Teacher behavior and student achievement. In M. C. Wittrock (Ed.), *Handbook of research on teaching* (3rd ed., pp. 328–375). New York: Macmillan.

Brown, J. S., & VanLehn, K. (1980). Repair theory: A generative theory of bugs in procedural skills. *Cognitive Science, 2*, 155–192.

Cai, J. (1987). *On mathematical abstracting ability and its training.* Unpublished master's thesis. Department of Mathematics, Beijing Normal University, Beijing, P. R. of China.

Cai, J. (1991, March). *Elementary school mathematics curriculum: A comparison between China and the U.S.* Paper presented at the 40th annual conference of the Pennsylvania Council of Teachers of Mathematics, Pittsburgh, PA.

Cai, J. (1995a). Beyond the computational algorithm: Students' understanding of the arithmetic aver-age concept. In L. Meira & D. Carraher (Eds.), *Proceedings of the Nineteenth Annual Meeting of the International Group for the Psychology of Mathematics Education* (Vol. 3, pp. 144–151). Recife, Brazil: Universidade Federal de Pernambuco.

Cai, J. (1995b). Exploring gender differences in solving open-ended mathematical problems. In D. T. Owens, M. K. Reed, and G. M. Millsaps (Eds.), *Proceedings of the Seventeenth Annual Meeting of the North American Chapter of the International Group of the Psychology of Mathematics Education* (Vol. 2, pp. 24–30). Columbus, OH: ERIC Clearinghouse for Science, Mathematics, and Environmental Education.

Cai, J., & Silver, E. A. (1993). *Mathematical abilities with symbol manipulation and mathematics understanding: A comparison between U.S. and China.* Paper presented at the annual meeting of the American Educational Research Association, Atlanta, GA.

Cai, J., & Silver, E. A. (1994). A cognitive analysis of Chinese students' mathematical problem solv-ing: An exploratory study. In D. Kirshner (Ed.), *Proceedings of the sixteenth annual meeting of North American Chapter of the International Group for the Psychology of Mathematics Educa-tion* (Vol. 2, pp. 3–9). Louisiana State University, Baton Rouge, Louisiana.

Cai, J., & Silver, E. A. (1995). Solution processes and interpretations of solutions in solving a divi-sion-with-remainder story problem: Do Chinese and U.S. students have similar difficulties? *Jour-nal for Research in Mathematics Education, 26,* 491–497.

Cai, J., Jakabcsin, S. M., & Lane, S. (in press-a). *The role of open-ended tasks and scoring rubrics in assessing students' mathematical reasoning and communication,* 1996 Yearbook of the National Council of Teachers of Mathematics. Reston, VA: NCTM.

Cai, J., Magone, M., Wang, N., & Lane, S. (in press-b). Describing student performance qualitative-ly: A way of thinking about assessment. *Mathematics Teaching in the Middle School.*

California State Department of Education (1989). *A question of thinking: A first look at students' performance on open-ended questions in mathematics.* Sacramento, CA: Author.

Cao, C., & Cai, J. (1989). *Mathematics pedagogy.* Nanjing, People's Republic of China: Jiangsu Educational Press.

Carpenter, T. P., Moser, J. M., & Romberg, T. (Eds.). (1982). *Addition and subtraction: A cognitive perspective.* Hillsdale, NJ: Erlbaum.

Carraher, T. N., Carraher, D. W., & Schliemann, A. D. (1987). Written and oral mathematics. *Jour-nal for Research in Mathematics Education, 18,* 287–318.

Charles, R., & Silver, E. A. (Eds.). (1988). *Research agenda for mathematics education: Teaching and assessing mathematical problem solving.* Reston, VA: NCTM (Co-published with Lawrence Erlbaum Associates, Hillsdale, NJ).

Chen, C., & Stevenson, H. W. (1989). Homework: A cross-cultural examination. *Child Development, 60,* 551–561.

Chi, M. T. H., Feltovich, P. J., & Glaser, R. (1981). Categorization and representation of physics problems by high-experienced subjects and low-experienced subjects. *Cognitive Psychology, 5,* 215–281.

Chi, M. T. H., Glaser, R., & Farr, M. J. (Eds.). (1988). *The nature of expertise.* Hillsdale, NJ: Erlbaum.

Cloeman, J. S., & Hoffer, T. (1987). Public and private high schools: *The impact of communities.* New York: Basic Books.

Cockcroft, W. H. (1982). *Mathematics counts: Report of the Committee of Inquiry Into the Teaching of Mathematics Under the Chairmanship of Dr. W. H. Cockcroft.* London: Her Majesty's Sta-tionery Office.

Cocking, R. R., & Mestre, J. P. (Eds.). (1988). *Linguistic and cultural influences on learning mathematics.* Hillsdale, NJ: Erlbaum.

Cogan, J., Torney-Purta, J., & Anderson, L. W. (1988). Knowledge and attitudes toward global issues: Students in Japan and the United States. *Comparative Education Review, 32*(3), 282–297.

Cookson, P. W. (1989). United States of America: Contours of continuity and controversy in private schools. In G. Walford (Ed.), *Private schools in ten countries: Policy and practice* (pp. 57–84). New York: Routledge.

Cronbach, L. J. (1951). Coefficient alpha and the internal structure of tests. *Psychometrika, 16,* 297–334.

Davis, R. B. (1986). Conceptual and procedural knowledge in mathematics: A summary analysis. In J. Hiebert (Eds.), *Conceptual and procedural knowledge: The case of mathematics* (pp. 265–300). Hillsdale, NJ: Erlbaum.

Division of Mathematics of People's Education Press. (1987). *National unified mathematics textbooks in elementary school.* Beijing: People's Education Press.

Duke, B. (1986). *The Japanese school: Lessons for industrial America.* New York: Praeger.

Eckstein, M. A., & Noah, H. J. (Eds.). (1991). *Examinations: Comparative and international studies.* New York: Pergamon.

Eicholz, R. E., O'Daffer, P. G., Charles, R. I., Young, S. L., & Barnett, C. S. (1987). *Addison-Wesley Mathematics (K–6).* Menlo Park, CA: Addison-Wesley.

Ericsson, K. A., & Simon, H. A. (1980). Verbal reports as data. *Psychological Review, 87*(3), 215–251.

Ericsson, K. A., & Simon, H. A. (1984). *Protocol analysis: Verbal reports as data.* Cambridge, MA: MIT Press.

Ernest, P. (1991). *The philosophy of mathematics education.* London: Falmer Press.

Farrell, J. P. (1979). The necessity of comparisons in the study of education: The salience of science and the problem of comparability. *Comparative Education Review, 23*(1), 3–16.

Flanders, J. R. (1987). How much of the content in mathematics textbooks is new? *Arithmetic Teacher, 35*(1), 18–23.

Frederick, W. C., & Walberg, H. J. (1980). Learning as a function of time. *Journal of Educational Research, 73*(4), 183–194.

Frederiksen, N., Glaser, R., Lesgold, A., & Shafto, M. G. (Eds.). (1990). *Diagnostic monitoring of skill and knowledge acquisition.* Hillsdale, NJ: Erlbaum.

Freedle, R. (Ed.). (1990). *Artificial intelligence and the future of testing.* Hillsdale, NJ: Erlbaum.

Fujita, H., Miwa, T., & Becker, J. P. (1990). The reform of mathematics education at the upper secondary school level in Japan. In I. Wirszup & R. Streit (Eds.), *Developments in school mathematics education around the world.* Reston, VA: NCTM.

Fuson, K. C., & Kwon, Y. (1991). Chinese-based regular and European irregular systems of number words: The disadvantages for English-speaking children. In K. Durkin & B. Shire (Eds.), *Language in mathematics education* (pp. 211–226). Philadelphia: Open University Press.

Fuson, K. C., & Kwon, Y. (1992). Learning addition and subtraction: Effects of number words and other cultural tools. In J. Bideau, C. Meljac, & J. P. Fischer (Eds.), *Pathways to number* (283–302). Hillsdale, NJ: Erlbaum.

Fuson, K. C., Stigler, J. W., & Bartsch, K. B. (1988). Grade placement of addition and subtraction topics in Japan, Mainland China, the Soviet Union, Taiwan, and the United States. *Journal for Research in Mathematics Education, 19,* 449–456.

Garden, R. A., & Livingstone, I. (1989). The contexts of mathematics education: Nations, communities and schools. In D. F. Robitaille, & R. A. Garden, *The IEA study of mathematics II: Contexts and outcomes of school mathematics* (pp. 17–38). New York: Pergamon.

Gifford, B. R., & O'Connor, M. C. (Eds.). (1992). *Changing assessments: Alternative views of aptitude, achievement and instruction.* Boston: Kluwer Academic Publishers.

Ginsburg, H. P. (Ed.) (1983). *The development of mathematical thinking.* New York: Academic Press.

Ginsburg, H. P., Choi, Y. E., Netley, R., Chi, C., Lopez, L., Song, M., Inagaki, K., & Kondo, H. (1990). *Early mathematical thinking: A comparative study of U.S., Chinese, Japanese and Korean children.* Paper presented at the annual meeting of the American Educational Research Association, Boston, MA.

Ginsburg, M., & Clift, R. (1990). The hidden curriculum of preservice teacher education. In W. R. Houston (Ed.), *Handbook of research on teacher education* (pp. 450–465). New York: Macmillan.

Glaser, R. (1987). The integration of instruction and testing. In D. C. Berliner & B. V. Rosenshine (Eds.), *Talks to teachers* (pp. 329–341). New York: Basic Books.

Glaser, R., Lesgold, A. M., & Lajoie, S. (1985). Toward a cognitive theory for the measurement of achievement. In R. R. Ronning, J. Glover, J. C. Conoley, & J. C. Witt (Eds.), *The influence of cognitive psychology on testing and measurement* (pp. 41–85). Hillsdale, NJ: Erlbaum.

Goldin, G. A. (1987). Cognitive representational systems for mathematical problem solving. In C. Janvier (Ed.), *Problems of representation in the teaching and learning of mathematical problem solving* (pp. 125–145). Hillsdale, NJ: Erlbaum.

Goldin, G. A. (1992). Toward an assessment framework for school mathematics. In R. Lesh & S. J. Lamon (Eds.), *Assessment of authentic performance in school mathematics* (pp. 64–89). AAAS Press.

Greeno, J. G., Riley, M. S., & Gelman, R. (1984). Conceptual competence and children's counting. *Cognitive Psychology, 16,* 94–143.

Grouws, D. A. (Ed.). (1992). *Handbook of research on mathematics teaching and learning.* New York: Macmillan.

Grouws, D. A., Cooney, T. J., & Jones, D. (Eds.). (1988). *Perspectives on research on effective mathematics teaching.* Hillsdale, NJ: Erlbaum.

Harnisch, D., Walberg, J., Tsai, S-L, Sato, T., & Fryans, L. (1985). Mathematics productivity in Japan and Illinois. *Evaluation in Education, 9,* 277–284.

Hashimoto, Y. (1987). Classroom practice of problem solving in Japanese elementary schools. In J. P. Becker & T. Miwa (Eds.), *Proceedings of the U. S.-Japan seminar on mathematical problem solving* (pp. 94–119). Carbondale, IL: Southern Illinois University.

Hatano, G. (1988). Social and motivational bases for mathematical understanding. In G. B. Saxe & M. Gearhart (Eds.), *Children's mathematics* (pp. 55–70). San Francisco: Jossey Bass.

Hatano, G. (1990). Toward the cultural psychology of mathematical cognition. In H. W. Stevenson & S. Lee (Eds.), *Contexts of achievement: A study of American, Chinese, and Japanese children* (pp. 108–115). Chicago: University of Chicago Press.

Henningsen, M., & Cai, J. (1993). *Achieving coherence in mathematics instruction using The One Problem-Multiple Changes Approach.* Paper presented at the annual meeting of the American Educational Research Association, Atlanta, GA.

Hess, R. D., Azuma, H., Kashigawa, K., Dickson, P., Nagano, K., Holloway, S., Miyake, K., Price, G., Hatano, G., & McDevitt, T. (1986). Family influences on school readiness and achievement in Japan and the United States: An overview of a longitudinal study. In H. Stevenson, H. Azuma, & K. Hakuta (Eds.), *Child development and education in Japan* (pp. 147–166). New York: Freeman.

Hess, R. D., & Azuma, H. (1991). Cultural support for schooling: Contrasts between Japan and the United States. *Educational Researcher, 20*(9), 2–8.

Hiebert, J. (Eds.). (1986). *Conceptual and procedural knowledge: The case of mathematics.* Hillsdale, NJ: Erlbaum.

Husen, T. (1967). *International study of achievement in mathematics: A comparison of twelve countries,* Volumes I & II. New York: John Wiley & Sons.

Ito-Hino, K. (1994). *Proportional reasoning and learning in American and Japanese sixth-grade students: Case studies.* Unpublished doctoral dissertation. Carbondale, IL: Southern Illinois University.

Jackson, P. W. (Ed.). (1992). *Handbook of research on curriculum.* New York: Macmillan.

James, T., & Levin, H. M. (Eds.). (1988). *Comparing public and private schools: Institutions and organizations.* New York: Falmer Press.

Keats, D. M. (1982). Cultural bases of concepts of intelligence: A Chinese versus Australian comparison. In P. Sukontasarp, N. Yongsiri, P. Intasuwan, N. Jotiban, & C. Suvannathat (Eds.), *Proceedings of the second Asian workshop on child and adolescent development* (pp. 67–75). Bangkok: Burapasilpa Press.

Kintsch, W., & Greeno, J. G. (1985). Understanding and solving word arithmetic problems. *Psychological Review, 92,* 109–129.

Kuhn, D. (1989). Children and adults as intuitive scientists. *Psychological Review, 96*(4), 674–689.

Kulm, G. (1994). *Mathematics assessment.* San Francisco: Jossey-Bass.

Lane, S. (1993). The conceptual framework for the development of a mathematics assessment for QUASAR. *Educational Measurement: Issues in Practice, 12*(2), 16–23.

Lane, S., & Silver, E. A. (in press). Equity and validity considerations in the design and implementation of a mathematics performance assessment: The experience of the QUASAR project. In M. T. Nettles (Ed.), *Equity and assessment in educational testing and assessment.* Boston, MA: Kluwer Academic Publishers.

Lane, S., Stone, C. A., Ankenmann, R. D., & Liu, M. (1994). Reliability and validity of a mathematics performance assessment. *International Journal of Educational Research, 21*(3), 247–266.

Lapointe, A. E., Mead, N. A., & Askew, J. M. (1992). *Learning mathematics.* Princeton, NJ: Educational Testing Service.

Lapointe, A. E., Mead, N. A., & Phillips, G. W. (1989). *A world of differences: An international assessment of mathematics and science*. Princeton, NJ: Educational Testing Service.

Leinhardt, G. (1982). Overlap: Testing whether it's taught. In G. F. Madaus (Ed.), *The courts, validity, and minimum competency testing*. Boston: Kluwer-Nijhoff.

Leinhardt, G. (1993). On teaching. In R. Glaser (Ed.), *Advances in instructional psychology* (Vol. 4, pp. 1–54). Hillsdale, NJ: Erlbaum.

Leinhardt, G., Putnam, R., & Hattrup, R. A. (Eds.). (1992). *Analyses of arithmetic for mathematics teaching*. Hillsdale, NJ: Erlbaum.

Lesgold, A., Lajoie, S., Logan, D., & Eggan, G. (1990). Applying cognitive task analysis and research methods to assessment. In N. Frederiksen, R. Glaser, A. Lesgold, & M. G. Shafto (Eds.), *Diagnostic monitoring of skill and knowledge acquisition* (pp. 325–350). Hillsdale, NJ: Erlbaum.

Lynch, E. W., & Hanson, M. J. (Eds.). (1992). *Developing cross-cultural competence: A guide for working with young children and their families*. Baltimore: Paul H. Brookes.

Lynn, R. (1982). IQ in Japan and the United States shows a growing disparity. *Nature, 297*, 22–223.

Lynn, R. (1988). *Educational achievement in Japan: Lessons for the West*. Armonk, NY: Sharpe.

Magone, M., Cai, J., Silver, E. A., & Wang, N. (1994). Validating the cognitive complexity and content quality of a mathematics performance assessment. *International Journal of Educational Research, 21*(3), 317–340.

Mathematics Sciences Education Board. (1993). *A conceptual guide for mathematics assessment*. Washington, DC: National Academy Press.

Mayer, R. E. (1987). *Educational psychology: A cognitive approach*. Boston: Little, Brown.

Mayer, R. E., Lewis, A. B., & Hegarty, M. (1992). Mathematical misunderstandings: Qualitative reasoning about quantitative problems. In J. I. D. Campbell (Ed.), *The nature and origins of mathematical skills* (pp. 137–154). Amsterdam: Elsevier Science Publishers.

Mayer, R. E., Tajika, H., & Stanley, C. (1991). Mathematical problem solving in Japan and the United States: A controlled comparison. *Journal of Educational Psychology, 83*(1), 69–72.

McLeod, D. B., & Adams, M. (Eds.). (1989). *Affect and mathematical problem solving: A new perspective*. New York: Springer-Verlag.

McKnight, C. C., Crosswhite, F. J., Dossey, J. A., Kifer, E., Swafford, J. O., Travers, K. J., & Cooney, T. J. (1987). *The underachieving curriculum: Assessing U.S. mathematics from an international perspective*. Champaign, IL: Stipes.

Medrich, E. A., & Griffith, J. E. (1992). *International mathematics and science assessments: What have we learned?* Washington, DC: Office of Educational Research and Improvement and National Center for Education Statistics, U.S. Department of Education (Report No. NCES92-011, 1992).

Messick, S. (1989). Validity. In R. L. Linn (Ed.), *Educational measurement* (3rd ed., pp. 13–104). New York: Macmillan.

Miller-Jones, D. (1989). Culture and testing. *American Psychologists, 44*(2), 360–366.

Miura, I. T. (1987). Mathematics achievement as a function of language. *Journal of Educational Psychology, 79*, 79–82.

Miura, I. T., & Okamoto, Y. (1989). Comparisons of U.S. and Japanese first graders' cognitive representation of number and understanding of place value. *Journal of Educational Psychology, 81*(1), 109–113.

Miura, I. T., & Okamoto, Y. (1993). *Solving arithmetic word problems: U.S. and Japanese comparisons*. Paper presented at the annual meeting of the American Educational Research Association, Atlanta, GA.

National Council of Teachers of Mathematics. (1989). *Curriculum and evaluation standards for school mathematics*. Reston, VA: Author.

National Council of Teachers of Mathematics. (1991). *Professional standards for teaching mathematics*. Reston, VA: Author.

National Council of Teachers of Mathematics. (1995). *Assessment standards for school mathematics*. Reston, VA: Author.

Newell, A., & Simon, H. A. (1972). *Human problem solving*. Englewood Cliffs, NJ: Prentice-Hall.

Niss, N. (Ed.). (1993). *Assessment in Mathematics Education and Its Effects*. London: Kluwer Academic Publishers.

Orr, E. W. (1987). *Twice as less: Black English and the performance of Black students in mathematics and science.* New York: W. W. Norton.

Pellegrino, J. M. (1992). Commentary: Understanding what we measure and measuring what we understand. In B. R. Gifford & M. C. O'Connor (Eds.), *Changing assessments: Alternative views of aptitude, achievement and instruction* (pp. 275–300). Boston: Kluwer Academic Publishers.

Polya, G. (1957). *How to solve it* (2nd ed.). New York: Doubleday.

Postlethwaite, T. N. (1987) Comparative education achievement research: Can it be improved? *Comparative Education Review, 31*(1), 150–158.

Postlethwaite, T. N. (1988). Preface. In T. N. Postlethwaite (Ed.), *Encyclopedia of comparative education and national systems of education* (pp. xvii–xxvi). Oxford: Pergamon.

Pressley, M., et al. (1990). *Cognitive strategy instruction that really improves children's academic performances.* New York: Brookline Books.

Putnam, R. T., Lampert, M., & Peterson, P. L. (1990). Alternative perspectives of knowing mathematics in elementary schools. *Review of Research in Education, 16,* 57–150.

Raivola, R. (1985). What is comparison? Methodological and philosophical considerations. *Comparative Education Review, 29*(3), 362–374.

Resnick, L. B. (1982). Syntax and semantics in learning to subtract. In T. P. Carpenter, J. M. Moser, & T. Romberg (Eds.), *Addition and subtraction: A cognitive perspective* (pp. 136–155). Hillsdale, NJ: Erlbaum.

Resnick, L. B. (1984). Beyond error analysis: The role of understanding in elementary school arithmetic. In H. Cheek (Ed.), *Diagnostic and prescriptive mathematics: Issues, ideas, and insight* (pp. 181–205). Kent, OH: Research Council for Diagnostic and Prescriptive Mathematics.

Resnick, L. B. (1987). Learning in school and out. *Educational Researcher, 16,* 13–19.

Resnick, L.B., Nesher, P., Leonard, F., Magone, M., Omanson, S., & Peled, I. (1989). Conceptual bases of arithmetic errors: The case of decimal fractions. *Journal for Research in Mathematics Education, 20,* 8–27.

Resnick, L. B., & Resnick, D. P. (1992). Assessing the thinking curriculum: New tools for educational reform. In B. R. Gifford & M. C. O'Connor (Eds.), *Changing assessments: Alternative views of aptitude, achievement and instruction* (pp. 37–76). Boston: Kluwer Academic Publishers.

Riley, M. S., Greeno, J. G., & Heller, J. I. (1993). Development of children's problem-solving ability in arithmetic. In H. P. Ginsburg (Ed.), *The development of mathematical thinking* (pp. 153–196). New York: Academic Press.

Robitaille, D. F. (1992). *Achievement measures in the Third International Mathematics and Science Study.* Paper presented at the annual meeting of the American Educational Research Association, San Francisco, CA.

Robitaille, D. F., & Garden, R. A. (1989). *The IEA study of mathematics II: Contexts and outcomes of school mathematics.* New York: Pergamon.

Robitaille, D. F., & Travers, K. J. (1992). International studies of achievement in mathematics. In D. A. Grouws (Ed.), *Handbook of research on mathematics teaching and learning* (pp. 687–709). New York: Macmillan.

Romberg, T. A., Zarinnia, E. A., & Collins, K. F. (1990). A new world view of assessment in mathematics. In. G. Kulm (Ed.), *Assessing higher order thinking in mathematics* (pp. 21–38). Washington, DC: American Association for the Advancement of Science.

Romberg, T. A., Wilson, L., Khaketla, M., & Chavarria, S. (1992). Curriculum and test alignment. In T. A. Romberg (Ed.), *Mathematics assessment and evaluation: Imperatives for mathematics educators* (pp. 61–74). State University of New York Press.

Ronning, R. R., Glover, J., Conoley, J. C., & Witt, J. C. (Eds.). (1985). *The influence of cognitive psychology on testing and measurement.* Hillsdale, NJ: Erlbaum.

Rotberg, I. C. (1990). I never promised you first place. *Phi Delta Kappan, 72,* 296–303.

Royer, J. M., Cisero, C. A., & Carlo, M. S. (1993). Techniques for assessing cognitive skills. *Review of Educational Research, 63*(2), 201–243.

Schoenfeld, A. H. (1979). Explicit heuristic training as a variable in problem-solving performance. *Journal for Research in Mathematics Education, 10,* 173–187.

Schoenfeld, A. H. (1985). *Mathematical problem solving.* Orlando, FL: Academic Press.

Segall, M. H., Dasen, P. R., Berry, J. W., & Poortinga, Y. H. (1990). *Human behavior in global perspective: An introduction to cross-cultural psychology.* New York: Pergamon.

Senk, S. L. (1985). How well do students write geometry proofs? *Mathematics Teacher, 78,* 448–456.

Senk, S. L. (1989). Van Hiele levels and achievement in writing geometry proofs. *Journal for Research in Mathematics Education, 20,* 309–321.

Shulman, L. S. (1986). Paradigms and research programs in the study of teaching: A contemporary perspective. In M. C. Wittrock (Ed.), *Handbook of research on teaching* (3rd ed., pp. 3–36). New York: Macmillan.

Siegler, R. S., & Shrager, J. (1984). Strategy choices in addition and subtraction: How do children know what to do? In C. Sophian (Ed.), *Origins of cognitive skills* (pp. 229–293). Hillsdale, NJ: Erlbaum.

Silver, E. A. (1987). Foundations of cognitive theory and research for mathematics problem solving. In A. H. Schoenfeld (Ed.), *Cognitive science and mathematics education* (pp. 33–60). Hillsdale, NJ: Erlbaum.

Silver, E. A. (1992). Assessment and mathematics education reform in the United States. *International Journal of Educational Research, 17*(5), 489–502.

Silver, E. A. (1993). *Quantitative Understanding: Amplifying Student Achievement and Reasoning.* Pittsburgh, PA: Learning Research and Development Center.

Silver, E. A. (1994). Mathematical thinking and reasoning for all students: Moving from rhetoric to reality. In D. F. Robitaille, D. H. Wheeler, & Kieran (Eds.), *Selected lectures from the 7th International Congress on Mathematics Education* (pp. 311–326). Sainte-Foy, Quebec: Les Presses De L'Université Laval.

Silver, E. A., & Cai, J. (1993). *Schemes for analyzing student responses to QUASAR's performance assessments: Blending cognitive and psychometric considerations.* Paper as part of a symposium presented at the annual meeting of the American Educational Research Association, Atlanta, GA.

Silver, E. A., & Lane, S. (1992). Assessment in the context of mathematics instruction reform: The design of assessment in the QUASAR project. In M. Niss (Ed.), *Assessment in mathematics education and its effects* (pp. 59–70). London: Kluwer Academic Publishers.

Silver, E. A., & Shapiro, L. J. (1992). Examinations of situation-based reasoning and sense-making in students' interpretations of solutions to a mathematical story problem. In J. P. Mendes da Ponte (Ed.), *Advances in mathematical problem solving* (pp. 113–123). Berlin: Springer-Verlag.

Silver, E. A., Kenney, P. A., & Salmon-Cox, L. (1992). The content and curricular validity of the 1990 NAEP mathematics items: A retrospective analysis. In G. Bohrnstedt (Ed.), *Assessing student achievement in the states: Background studies* (pp. 157–218). Stanford, CA: Stanford University, National Academy of Education.

Silver, E. A., Leung, S. S., & Cai, J. (1995). Generating multiple solutions for a problem: A comparison of the responses of U.S. and Japanese students. *Educational Studies in Mathematics, 28*(1), 35–54.

Simon, H. A. (1979). *Models of thought* (Vol. I). New Haven: Yale University Press.

Simon, H. A. (1989). *Models of thought* (Vol. II). New Haven: Yale University Press.

Sleeman, D., Kelly, A. E., Martinak, R., Ward, R. D., & Moore, J. L. (1989). Studies of diagnosis and remediation with high school algebra students. *Cognitive Science, 13,* 551–568.

Snow, R. E., & Lohman, D. F. (1989). Implications of cognitive psychology for educational measurement. In R. L. Linn (Ed.), *Educational measurement* (3rd ed. pp. 263–332). New York: Macmillan.

Song, M. J., & Ginsburg, H. P. (1987). The development of informal and formal mathematical thinking in Korean and U.S. children. *Child Development, 58,* 1286–1296.

State Education Commission of China (1987). *Mathematical syllabus in elementary school.* Beijing: People's Education Press.

Steinberg, R. M., Sleeman, D. H., & Ktorza, D. (1991). Algebra students' knowledge of equivalence of equations. *Journal for Research in Mathematics Education, 22,* 112–121.

Sternberg, R. J. (Ed.). (1982). *Handbook of human intelligence.* Cambridge: Cambridge University Press.

Sternberg, R. J. (1984). What cognitive psychology can (and cannot) do for test development. In B. S. Plake (Ed.), *Social and technical issues in testing: Implications for test construction and usage* (pp. 39–60). Hillsdale, NJ: Erlbaum.

Sternberg, R. J. (1991). Cognitive theory and psychometrics. In R. K. Hambleton & J. N. Zaal (Eds.), *Advances in educational and psychological testing: Theory and applications* (pp. 367–393). Boston: Kluwer Academic Publishers.

Stevens, J. (1992). *Applied multivariate statistics for the social sciences* (2nd ed.). Hillsdale, NJ: Erlbaum.

Stevenson, H. W., & Azuma, H. (1983). IQ in Japan and the United States: Methodological problems in Lynn's analysis. *Nature, 306*, 291–292.

Stevenson, H. W., & Bartsch, (1992). An analysis of Japanese and American textbooks in mathematics. In R. Leetsman & H. J. Walberg (Eds.), *Japanese educational productivity* (pp. 103–134). Greenwich, CT: JAI Press.

Stevenson, H. W., Chen, C., & Lee, S. (1993). Mathematical achievement of Chinese, Japanese, and American children: Ten years later. *Science, 259*, 53–58.

Stevenson, H.W., & Lee, S. (1990). *Contexts of achievement: A study of American, Chinese, and Japanese children.* Chicago, IL: University of Chicago Press.

Stevenson, H. W. Lee, S., Chen, C., Lummis, M., Stigler, J. W., Liu, F., & Fang, G. (1990). Mathematics achievement of children in China and the United States. *Child Development, 61*, 1053–1066.

Stevenson, H. W., Lee, S., & Stigler, J. W. (1986). Mathematics achievement of Chinese, Japanese, and American children. *Science, 231*, 693–699.

Stevenson, H. W., & Stigler, J. W. (1992). *The learning gap.* New York: Summit.

Stevenson, H. W., Stigler, J. W., Lee, S., Lucker, W., Kitamura, S., & Hsu, C. (1985). Cognitive performance and academic achievement Japanese, Chinese, and American children. *Child Development, 56*, 718–734.

Stigler, J. W., & Fernandez, C. (1993). *Learning mathematics from classroom instruction: Cross-cultural and experimental perspectives.* Paper presented at the Minnesota Symposium on Child Psychology, Minneapolis, MN.

Stigler J. W., Lee, S. Y., Lucker, G. W., & Stevenson, H. W. (1982): Curriculum and achievement in mathematics: A study of elementary school children in Japan, Taiwan, and the United States. *Journal of Educational Psychology, 74*, 315–322.

Stigler J. W., Lee, S., & Stevenson, H. W. (1987). Mathematics classrooms in Japan, Taiwan, and the United States. *Child Development, 58*, 1272–1285.

Stigler J. W., Lee, S., & Stevenson, H. W. (1990). *Mathematical knowledge of Japanese, Chinese, and American elementary school children.* Reston, VA: NCTM.

Stigler, J. W., & Perry, M. (1988). Cross-cultural studies of mathematics teaching and learning: Recent findings and new directions. In D. A. Grouws, T. J. Cooney, & D. Jones (Eds.), *Effective mathematics teaching* (pp. 104–223). Reston, VA: NCTM.

Stigler, J. W., & Stevenson, H. W. (Spring, 1991). How Asian teachers polish each lesson to perfection. *American Educator*, 12–47.

Tall, D. (Ed.). (1991). *Advanced mathematical thinking.* Dordrecht, Netherlands: Kluwer Academic Publishers.

Theisen, G. L., Achola, P. W, & Boakari, F. M. (1983). The underachievement of cross-national studies of achievement. *Comparative Education Review, 27*(1), 46–48.

Tian, W., & et al. (1989). *Report on the first national wide investigation of mathematics teaching in the third-grade of junior high school in China.* Department of Mathematics, East China Normal University.

Travers, K. J., & Westbury, I. (1989). *The IEA study of mathematics I: Analysis of mathematics curricula.* New York: Pergamon.

Voss, J. F., Perkins, D. N., & Segal, J. W. (Eds.). (1991). *Informal reasoning and education.* Hillsdale, NJ: Erlbaum.

Walberg, H. J. (1988). Synthesis of research on time and learning. *Educational Leadership, 45*(6), 76–85.

Wang, M. C., Haertel, G. D., & Walberg, H. J. (1993). Toward a knowledge base for school learning. *Review of Educational Research, 63*(3), 249–294.

Weinberg, R. A. (1989). Intelligence and IQ: Landmark issues and great debates. *American Psychologists, 44*(2), 98–104.

Westbury, I. (1992). Comparing American and Japanese achievement: Is the United States really a low achiever? *Educational Researcher, 21*(5), 18–24.

Westbury, I., Ethington, C. A., Sosniak, L. A., Baker, D. P. (1994). *In search of more effective mathematics education.* Norwood, NJ: Ablex Publishing Corporation.

Wittrock, M. C. (1990). Testing and recent research in cognition. In M. C. Wittrock & E. L. Baker (Eds.), *Testing and cognition* (pp. 5–16). Prentice Hall.

Wittrock, M. C. (Ed.). (1992). *Handbook of research on teaching* (3rd ed.). New York: Macmillan.

Wittrock, M. C., & Baker, E. L. (Eds.). (1991). *Testing and cognition.* Prentice Hall.

Zhong, S. (1988). On mathematical problem solving: A conversation with students. *Education in School Subjects, 2,* 1–3.

Appendix A
Computation Tasks

Name: _____
                first   middle   last

Birthday: _____
                month   day   year you were born

School: _____ **Grade:** _____

Circle a or b: a. male   b. female

Instructions

This test has 20 math problems for you to solve. For each problem there will be four possible answers labeled a, b, c, and d. Your job is to circle the letter next to the correct answer. Here's a sample problem for you to try. Circle one of the letters.

$$5 \times 3 =$$

a. $\dfrac{3}{5}$

b. 2

c. 8

d. 15

The correct answer is 15 so you should circle the letter d.

When I say "START" you should turn the page and begin working on the problems. You will have 20 minutes. If you finish one page go on to the next page. Keep working until I tell you to stop. Check your answers if you finish early.

DO NOT TURN THIS PAGE UNTIL I SAY "START"

(1)

$$5.3 - 4.6 =$$

a.   .3
b.   .7
c.   1.3
d.   1.7

(2)

$$46\overline{)3572}$$

a.   77
b.   77R30
c.   78
d.   78R30

(3)

$$\begin{array}{r} 38.15 \\ -9.43 \\ \hline \end{array}$$

a.   28.72
b.   29.72
c.   31.32
d.   38.72

(4)

$$3600\overline{)828000}$$

a.   23
b.   230
c.   2300
d.   23000

(5)

$$12\overline{)13.08}$$

a.   1.09
b.   1.9
c.   10.9
d.   19

GO ON TO THE NEXT PAGE

(6)

$$0.034$$
$$\times \ 17$$

a.   .00578

b.   .0578

c.   .578

d.   5.78

---

(7)

$$5 + \left(-4\right) =$$

a.   −1

b.   1

c.   9

d.   −9

---

(8)

$$\frac{3}{4} - \frac{1}{6} =$$

a. $\dfrac{2}{24}$

b. $\dfrac{2}{6}$

c. $\dfrac{7}{12}$

d. $\dfrac{2}{2}$

---

(9)

$$0.08 \times 10 =$$

a.   .08

b.   .8

c.   8

d.   80

---

(10)

$$\frac{3}{8} \div 4$$

a. $\dfrac{3}{32}$

b. $\dfrac{12}{32}$

c. $\dfrac{3}{2}$

d. $\dfrac{12}{8}$

GO ON TO THE NEXT PAGE

(11)

$$\frac{3}{5} \times \frac{1}{9} =$$

a. $\frac{3}{45}$

b. $\frac{3}{14}$

c. $\frac{2}{4}$

d. $\frac{27}{5}$

(12)

$$6 \times \frac{4}{7} =$$

a. $\frac{4}{42}$

b. $\frac{24}{42}$

c. $\frac{24}{7}$

d. $6\frac{4}{7}$

(13)

$$\frac{5}{11} \div \frac{1}{9} =$$

a. $\frac{5}{99}$

b. $\frac{6}{20}$

c. $\frac{11}{45}$

d. $\frac{45}{11}$

(14)

$$\frac{5}{6} + \frac{3}{4} =$$

a. $\frac{8}{24}$

b. $\frac{8}{12}$

c. $\frac{8}{10}$

d. $1\frac{7}{12}$

(15)

$$1.5 + 9 + .3 =$$

a. 2.7

b. 10.8

c. 12.6

d. 16.2

GO ON TO THE NEXT PAGE

$\left(16\right)$

$$\frac{7}{10} = \frac{21}{\square}$$

The number in the $\square$ should be:

   a.   10

   b.   30

   c.   70

   d.   147

$\left(17\right)$

$$0.025\overline{)36}$$

   a.   1.44

   b.   14.4

   c.   144

   d.   1440

$\left(18\right)$

$$9\frac{1}{2} - 7\frac{2}{5} =$$

   a. $2\frac{9}{10}$

   b. $2\frac{2}{5}$

   c. $2\frac{3}{10}$

   d. $2\frac{1}{10}$

$\left(19\right)$

$$\left[\left(5\times3\right)+\left(6\times4\right)\right]\div3 =$$

   a.   13

   b.   23

   c.   37

   d.   42

$\left(20\right)$

$$\frac{\square}{5} = \frac{16}{20}$$

The number in the $\square$ should be:

   a. $\frac{1}{4}$

   b.   2

   c.   4

   d.   10

END

Appendix B

Component Questions

Name: _____

Instructions

This test has 18 math questions for you to answer. For each question there will be four possible answers labeled a, b, c, and d. Your job is to circle the letter next to the correct answer. The questions do not ask you to compute anything, so you should not do any arithmetic. Here's a sample problem to try. Circle a letter.

Which numbers are needed to solve this problem?

Marbles come in bags of 5 marbles each and each bag costs 25 cents. You want to buy 10 marbles. How many bags of marbles should you buy?

   a. 5, 25, 10
   b. 5, 25
   c. 5, 10
   d. 10

You need to use only 5 and 10 so you should circle letter c. Now try this problem.

Which operations should you carry out to solve this problem?

There are 12 hats and 24 children. How many children will not get hats?

   a. add, then subtract
   b. divide, then subtract
   c. divide only
   d. subtract only

The correct answer is to subtract 12 from 24 so you should circle d. Now try this.

Which number sentence is correct?

Apples come in crates of 72 apples each. There are 6 crates.

   a. the total number of apples = 72×6
   b. the total number of apples $\times 6 = 72$
   c. the total number of apples $\times 72 = 6$
   d. the total number of apples = 72

You should circle the letter a. If you multiply the number of apples in each crate (72) by the number of crates (6) times you will find the total number of apples.

Remember that you never have to compute a solution; just answer the question. When I say "START" you should turn the page and begin working on the problems. You will have 15 minutes. If you finish one page go on to the next page. Keep working until I tell you to stop. Check over your answers if you finish early.

DO NOT TURN THIS PAGE UNTIL I SAY "START"

(1) Which numbers are needed to solve this problem?
A package of 5 pencils costs 59 cents. Richie bought 3 packages and gave the cashier $2. How many pencils did he buy?

    a. 5, 59, 3, 2
    b. 59, 3, 2
    c. 5, 59, 3
    d. 5,3

(2) Which operations should you carry out to solve this problem?
The 200 children at a school are going on a bus trip. Each bus holds 50 children. How many buses are needed?

    a. divide, then add
    b. subtract only
    c. multiply only
    d. divide only

(3) Which number sentence is correct?
Ann and Rose have 20 books altogether.

    a. Ann's books = Rose's books + 20
    b. Ann's books + 20 = Rose's books
    c. Ann's books + Rose's books = 20
    d. Ann's books = Rose's books

(4) Which numbers are needed to solve this problem?
Karin's home is 8 blocks from her school. School starts at 8:00. She left home at 7:42 and arrived at school at 7:54. How long did it take her to get there?

    a. 8, 8:00, 7:42, 7:54
    b. 8:00, 7:42, 7:54
    c. 8:00, 7:54
    d. 7:42, 7:54

(5) Which operations should you carry out to solve this problem?
There are 30 students in a class, including 12 boys and 18 girls. The teacher asks them to get into groups of 3. How many groups are there?

    a. add, then multiply
    b. divide, then divide
    c. divide only
    d. subtract only

(6) Which number sentence is correct?
Dave and Sharon ate 12 candies altogether.

    a. number of candies Dave ate = number of candies Sharon ate + 12
    b. number of candies Dave ate + 12 = number of candies Sharon ate
    c. number of candies Dave ate + number of candies Sharon ate = 12
    d. number of candies Dave ate = number of candies Sharon ate

GO ON TO THE NEXT PAGE

(7) Which numbers are needed to solve this problem?
Lucia had $3 for lunch. She bought a sandwich for $.95, an apple for $.20, and milk for $.45. How much money did she spend?

    a. 3, 0.95, 0.20, 0.45
    b. 0.95, 0.20, 0.45
    c. 0.95, 0.45
    d. 3

(8) Which operations should you carry out to solve this problem?
Twelve candies come in each bag at the store. You buy 3 bags on Monday, 2 bags on Wednesday, and 1 bag on Friday. How many candies do you have?

    a. add, then multiply
    b. add, then divide
    c. add only
    d. divide only

(9) Which number sentence is correct?
John has 5 more marbles than Pete.

    a. John's marbles = 5 + Pete's marbles
    b. John's marbles + 5 = Pete's marbles
    c. John's marbles + Pete's marbles = 5
    d. John's marbles = 5

(10) Which numbers are needed to solve this problem?
Recess at Mountain View School starts at 10:00 and is over at 10:20. Lunch starts at 12:15. If it is 9:40 right now,. how many minutes are there before recess?

    a. 10:00, 10:20, 12:15, 9:40
    b. 10:00, 12:15
    c. 10:00, 10:20, 12:15
    d. 10:00, 9:40

(11) Which operations should you carry out to solve this problem?
If it costs 50 cents per hour to rent roller skates, what is the cost of using the skates from 1:00 p.m. to 3:00 p.m.?

    a. subtract, then multiply
    b. subtract, then divide
    c. add, then divide
    d. multiply only

(12) Which number sentence is correct?
Charlie's dog weighs 6 more pounds than Mario's dog.

    a. the weight of Charlie's dog = 6 + weight of Mario's dog
    b. the weight of Charlie's dog + 6 = weight of Mario's dog
    c. the weight of Charlie's dog + weight of Mario's dog = 6
    d. the weight of Charlie's dog = 6

GO ON TO THE NEXT PAGE

(13) Which numbers are needed to solve this problem?
It takes Kelly 15 minutes to walk 3 blocks to school. Kenny lives 4 blocks from school and he needs 5 more minutes than Kelly to walk to school. How long does it take for Kenny to walk to school?

- a. 15, 3, 5, 4
- b. 15, 5, 4
- c. 15,3
- d. 15, 5

(14) Which operations should you carry out to solve this problem?
You need to bring enough cookies so everyone at the class party can have 2 cookies each. There are 20 people at the party. Cookies come in boxes of 10 cookies each. How many boxes should you bring?

- a. divide, then add
- b. multiply, then divide
- c. divide only
- d. multiply only

(15) Which number sentence is correct?
Sally is 12 years old. This is 3 years older than Maria.

- a. Sally's age + 3 = Maria's age
- b. Sally's age = Maria's age + 3
- c. Sally's age + Maria's age = 3
- d. Maria's age = 12 + 3

(16) Which numbers are needed to solve this problem?
Mike spent $8 for 2 packages of large nails. Donald spent $4 more than Mike and bought 6 packages of small nails. How much did Donald spend?

- a. 8,2,4,6
- b. 8,4,6
- c. 8,4
- d. 4

(17) Which operations should you carry out to solve this problem?
On five tests in your math class your scores are 98, 63, 72, 86, and 100. What is your average score?

- a. add, then multiply
- b. add, then divide
- c. divide only
- d. multiply, then subtract

(18) Which number sentence is correct?
An apple costs 10 cents. This is 5 cents more than the cost of a banana.

- a. cost of an apple = cost of a banana + 5
- b. cost of an apple + 5 = cost of a banana
- c. cost of an apple + cost of a banana = 5
- d. cost of a banana = 10 + 5

END

Appendix C

Five of the Open-Ended Problems

(1) Students and teachers at Miller Elementary School will go by bus to have Spring sightseeing. There is a total of 296 students and teachers. Each bus holds 24 people.

How many buses are needed?

Show your work.

Explain your answer.

Answer:_____

(2) The shaded region below represents an island.

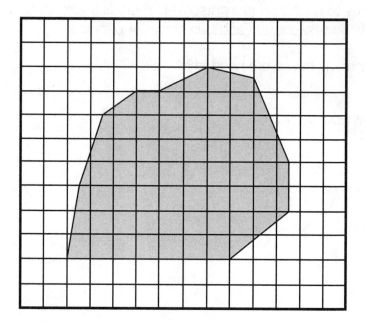

Each small square equals one square mile.

A. *Estimate* the area of the island in square miles.

Answer: _____

B. Explain how you found your estimate. You may use the drawing above in your explanation.

(3) Angela is selling hats for the Mathematics Club. This picture shows the number of hats Angela sold during the first three weeks.

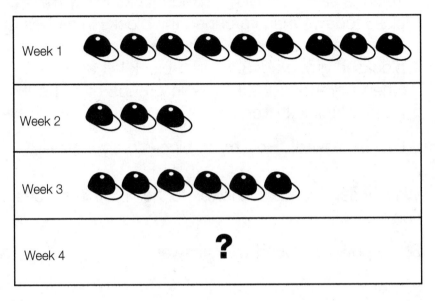

How many hats must Angela sell in Week 4 so that the *average* number of hats sold is 7?

Show how you found your answer.

Answer: _____

(4) Yolanda was telling her brother Damian about what she did in math class.

Yolanda said, "Damian, I used blocks in my math class today. When I grouped the blocks in groups of 2, I had 1 block left over. When I grouped the blocks in groups of 3, I had 1 block left over. And when I grouped the blocks in groups of 4, I still had 1 block left over."

Damian asked, "How many blocks did you have?"

What was Yolanda's answer to her brother's question?

Show how you found your answer.

Answer: _____

(5) Look at the pattern below.

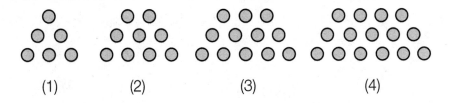

(1) (2) (3) (4)

A. Draw the fifth figure.

B. Draw the seventh figure.

C. Describe how you knew what the 7th figure would look like.

Appendix D

Teacher Questionnaire for the Booklet of Open-Ended Problems

School: _____

Questionnaire for Teachers

Dear teachers:

I would like to take a few minutes to ask you the following two questions. Please answer each question by choosing one of the three given choices.

1. Do you think your students are familiar with the task format in this booklet?

      a. YES   b. NOT SURE   c. NO

2. Have your sixth-grade students been taught enough information to answer each of the tasks in this booklet correctly?

Problem (1)

      a. YES   b. NOT SURE   c. NO

Problem (2)

      a. YES   b. NOT SURE   c. NO

Problem (3)

      a. YES   b. NOT SURE   c. NO

Problem (4)

      a. YES   b. NOT SURE   c. NO

Problem (5)

      a. YES   b. NOT SURE   c. NO

Problem (6)

      a. YES   b. NOT SURE   c. NO

Problem (7)

      a. YES   b. NOT SURE   c. NO

Appendix E

Item Analysis of Computation Tasks

Item	Alternatives		Percent of Students	
			U.S.	China
(1) $5.3 - 4.6 =$	a.	.3	2	1
	b.*	.7	89	99
	c.	1.3	7	1
	d.	1.7	2	0
(2) $46\overline{)3572}$	a.	77	4	2
	b.*	77R30	86	96
	c.	78	3	1
	d.	78R30	5	1
	Omitted		2	0
(3) $\begin{array}{r} 38.15 \\ -9.43 \end{array}$	a.*	28.72	91	94
	b.	29.72	6	5
	c.	31.32	1	0
	d.	38.72	2	1
(4) $3600\overline{)828000}$	a.	23	16	8
	b.*	230	65	90
	c.	2300	7	1
	d.	23000	10	1
	Omitted		2	0
(5) $12\overline{)13.08}$	a.*	1.09	86	94
	b.	1.9	7	2
	c.	10.95	5	3
	d.	19	1	1
	Omitted		1	0

Item	Alternatives		Percent of Students U.S.	Percent of Students China
(6) $\begin{aligned} 0.034 \\ \times\ 17 \end{aligned}$	a.	.00578	9	1
	b.	.0578	4	3
	c.*	.578	83	95
	d.	5.78	3	1
	Omitted		1	14
(7) $5+\left(-4\right)=$	a.	−1	11	8
	b.*	1	68	42
	c.	9	6	6
	d.	−9	15	30
	Omitted		1	14
(8) $\dfrac{3}{4}-\dfrac{1}{6}=$	a.	$\dfrac{2}{24}$	4	1
	b.	$\dfrac{2}{6}$	15	1
	c.*	$\dfrac{7}{12}$	54	98
	d.	$\dfrac{2}{2}$	25	0
	Omitted		2	0
(9) $0.08\times10=$	a.	.08	20	1
	b.*	.8	60	98
	c.	8	6	1
	d.	80	14	0
	Omitted		1	0
(10) $\dfrac{3}{8}\div4$	a.*	$\dfrac{3}{32}$	34	93
	b.	$\dfrac{12}{32}$	18	1
	c.	$\dfrac{3}{2}$	32	5
	d.	$\dfrac{12}{8}$	13	1
	Omitted		3	0

Item	Alternatives		Percent of Students	
			U.S.	China
(11) $\dfrac{3}{5} \times \dfrac{1}{9} =$	a.*	$\frac{3}{45}$	83	97
	b.	$\frac{3}{14}$	6	1
	c.	$\frac{2}{4}$	1	1
	d.	$\frac{27}{5}$	9	0
	Omitted		2	0
(12) $6 \times \dfrac{4}{7} =$	a.	$\frac{4}{42}$	4	5
	b.	$\frac{24}{42}$	44	0
	c.*	$\frac{24}{7}$	46	93
	d.	$6\frac{4}{7}$	5	1
	Omitted		1	0
(13) $\dfrac{5}{11} \div \dfrac{1}{9} =$	a.	$\frac{5}{99}$	40	2
	b.	$\frac{6}{20}$	5	0
	c.	$\frac{11}{45}$	18	1
	d.*	$\frac{45}{11}$	32	96
	Omitted		5	0
(14) $\dfrac{5}{6} + \dfrac{3}{4} =$	a.	$\frac{8}{24}$	7	1
	b.	$\frac{8}{12}$	7	2
	c.	$\frac{8}{10}$	47	2
	d.*	$1\frac{7}{12}$	37	95
	Omitted		2	0
(15) $1.5 + 9 + .3 =$	a.	2.7	26	10
	b.*	10.8	68	89
	c.	12.6	2	0
	d.	16.2	1	1
	Omitted		2	0

Item	Alternatives		Percent of Students	
			U.S.	China
(16) $\dfrac{7}{10} = \dfrac{21}{\square}$ The number in the $\square$ should be:	a.	10	7	1
	b.	30	87	94
	c.	70	5	4
	d.	147	0	1
	Omitted		1	0
(17) $0.025\overline{)36}$	a.	1.44	27	3
	b.	14.4	17	1
	c.	144	14	5
	d.	1440	36	91
	Omitted		6	0
(18) $9\dfrac{1}{2} - 7\dfrac{2}{5} =$	a.	$2\frac{9}{10}$	6	0
	b.	$2\frac{2}{5}$	16	1
	c.	$2\frac{3}{10}$	12	1
	d.	$2\frac{1}{10}$	61	98
	Omitted		5	0
(19) $\left[(5\times 3) + (6\times 4) \right] \div 3 =$	a.	13	90	98
	b.	23	0	1
	c.	37	2	1
	d.	42	3	0
	Omitted		4	0
(20) $\dfrac{\square}{5} = \dfrac{16}{20}$ The number in the $\square$ should be:	a.	$\frac{1}{4}$	0	1
	b.	2	4	2
	c.	4	90	96
	d.	10	2	1
	Omitted		3	0

Appendix F

Item Analysis of Component Questions

| | Percent of Students | |
Item	U.S.	China

(1) Which numbers are needed to solve this problem?

A package of 5 pencils costs 59 cents. Richie bought 3 packages and gave the cashier $2. How many pencils did he buy?

	U.S.	China
a. 5, 59, 3, 2	8	10
b. 59, 3, 2	12	14
c. 5, 59, 3	9	10
d.* 5, 3	70	65
Omitted	1	0

(2) Which operations should you carry out to solve this problem?

The 200 children at a school are going on a bus trip. Each bus holds 50 children. How many buses are needed?

	U.S.	China
a. divide, then add	2	0
b. subtract only	3	1
c. multiply only	9	2
d.* divide only	86	96
Omitted	0	0

(3) Which number sentence is correct?

Ann and Rose have 20 books altogether.

	U.S.	China
a. Ann's books = Rose's books + 20	3	2
b. Ann's books + 20 = Rose's books	0	2
c.* Ann's books + Rose's books = 20	93	91
d. Ann's books = Rose's books	4	5
Omitted	0	0

(4) Which numbers are needed to solve this problem?

Karin's home is 8 blocks from her school. School starts at 8:00. She left home at 7:42 and arrived at school at 7:54. How long did it take her to get there?

	U.S.	China
a. 8, 8:00, 7:42, 7:54	2	3
b. 8:00, 7:42, 7:54	10	7
c. 8:00, 7:54	1	4
d.* 7:42, 7:54	85	86
Omitted	1	0

| | Percent of Students | |
Item	U.S.	China

(5) Which operations should you carry out to solve this problem?

There are 30 students in a class, including 12 boys and 18 girls. The teacher asks them to get into groups of 3. How many groups are there?

	U.S.	China
a. add, then multiply	9	6
b. divide, then divide	6	5
c.* divide only	81	87
d. subtract only	4	2
Omitted	0	0

(6) Which number sentence is correct?

Dave and Sharon ate 12 candies altogether.

	U.S.	China
a. number of candies Dave ate = number of candies Sharon ate + 12	2	2
b. number of candies Dave ate + 12 = number of candies Sharon ate	3	2
c.* number of candies Dave ate + number of candies Sharon ate = 12	91	91
d. number of candies Dave ate = number of candies Sharon ate	4	5
Omitted	0	0

(7) Which numbers are needed to solve this problem?

Lucia had $3 for lunch. She bought a sandwich for $.95, an apple for $.20, and milk for $0.45. How much money did she spend?

	U.S.	China
a. 3, .95, .20, .45	16	17
b.* .95, .20, .45	82	81
c. .95, .45	2	1
d. 3	0	1
Omitted	0	0

(8) Which operations should you carry out to solve this problem?

Twelve candies come in each bag at the store. You buy 3 bags on Monday, 2 bags on Wednesday, and 1 bag on Friday. How many candies do you have?

	U.S.	China
a.* add, then multiply	59	77
b. add, then divide	7	7
c. add only	32	16
d. divide only	4	1
Omitted	0	1

| | Percent of Students | |
Item	U.S.	China

(9) Which number sentence is correct?

John has 5 more marbles than Pete.

a.* John's marbles = 5 + Pete's marbles	47	82
b. John's marbles + 5 = Pete's marbles	27	12
c. John's marbles + Pete's marbles = 5	6	3
d. John's marbles = 5	20	2
Omitted	0	1

(10) Which numbers are needed to solve this problem?

Recess at Mountain View School starts at 10:00 and is over at 10:20. Lunch starts at 12:15. If it is 9:40 right now, how many minutes are there before recess?

a. 10:00, 10:20, 12:15, 9:40	6	6
b. 10:00, 12:15	4	1
c. 10:00, 10:20, 12:15	4	5
d.* 10:00, 9:40	86	86
Omitted	0	1

(11) Which operations should you carry out to solve this problem?

If it costs 50 cents per hour to rent roller skates, what is the cost of using the skates from 1:00 p.m. to 3:00 p.m.?

a.* subtract, then multiply	38	53
b. subtract, then divide	14	4
c. add, then divide	14	2
d. multiply only	34	40
Omitted	0	1

(12) Which number sentence is correct?

Charlie's dog weighs 6 more pounds than Mario's dog.

a.* the weight of Charlie's dog = 6 + weight of Mario's dog	53	86
b. the weight of Charlie's dog + 6 = weight of Mario's dog	24	10
c. the weight of Charlie's dog + weight of Mario's dog = 6	6	2
d. the weight of Charlie's dog = 6	17	0
Omitted	1	1

	Percent of Students	
Item	U.S.	China

(13) Which numbers are needed to solve this problem?

It takes Kelly 15 minutes to walk 3 blocks to school. Kenny lives 4 blocks from school and he needs 5 more minutes than Kelly to walk to school. How long does it take for Kenny to walk to school?

	U.S.	China
a. 15, 3, 5, 4	16	6
b. 15, 5, 4	13	9
c. 15, 3	3	5
d.* 15, 5	67	81
Omitted	1	1

(14) Which operations should you carry out to solve this problem?

You need to bring enough cookies so everyone at the class party can have 2 cookies each. There are 20 people at the party. Cookies come in boxes of 10 cookies each. How many boxes should you bring?

	U.S.	China
a. divide, then add	15	5
b.* multiply, then divide	44	57
c. divide only	12	22
d. multiply only	28	12
Omitted	1	4

(15) Which number sentence is correct?

Sally is 12 years old. This is 3 years older than Maria.

	U.S.	China
a. Sally's age + 3 = Maria's age	28	9
b.* Sally's age = Maria's age + 3	59	77
c. Sally's age + Maria's age = 3	4	4
d. Maria's age = 12+3	8	8
Omitted	0	1

(16) Which numbers are needed to solve this problem?
Mike spent $8 for 2 packages of large nails. Donald spent $4 more than Mike and bought 6 packages of small nails. How much did Donald spend?

	U.S.	China
a. 8, 2, 4, 6	11	8
b. 8, 4, 6	15	9
c.* 8, 4	70	78
d. 4	3	2
Omitted	0	2

| | Percent of Students | |
Item	U.S.	China

(17) Which operations should you carry out to solve this problem?
On five tests in your math class your scores are 98, 63, 72, 86, and 100. What is your average score?

	U.S.	China
a. add, then multiply	4	2
b.* add, then divide	88	92
c. divide only	5	2
d. multiply, then subtract	2	1
Omitted	0	4

(18) Which number sentence is correct?
An apple costs 10 cents. This is 5 cents more than the cost of a banana.

	U.S.	China
a.* cost of an apple = cost of a banana + 5	62	75
b. cost of an apple + 5 = cost of a banana	27	11
c. cost of an apple + cost of a banana = 5	4	1
d. cost of a banana = 10 + 5	5	8
Omitted	1	4

Appendix G

Percentage Distributions of U.S. and Chinese Students at Each
Score Level for Each of the Open-Ended Problems

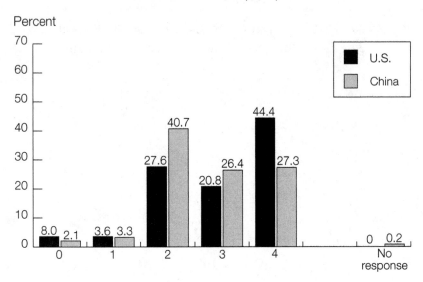

Division Problem (OE-1)

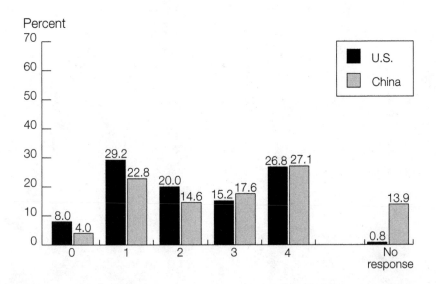

Estimation Problem (OE-2)

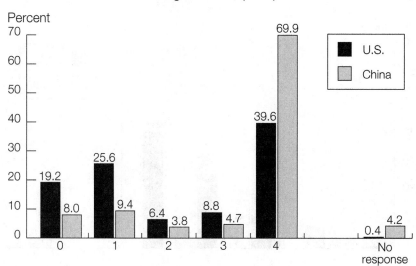

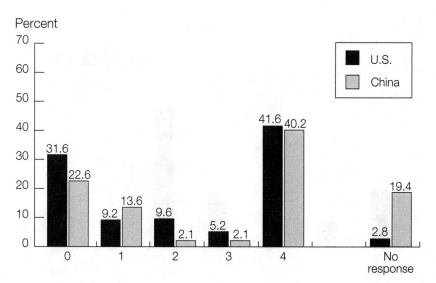

Pattern Problem (OE-5)

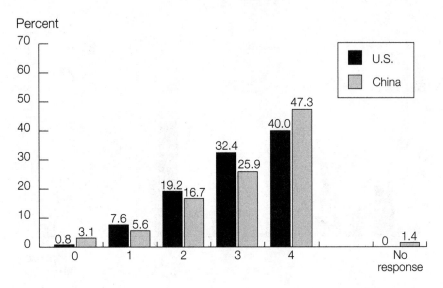

Ratio and Proportion Problem (OE-6)

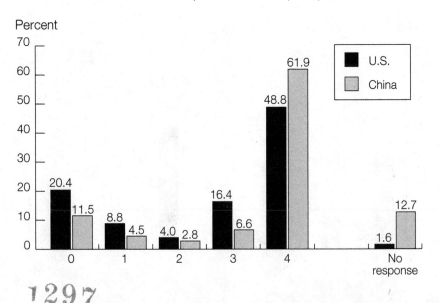

Prealgebra Problem (OE-7)

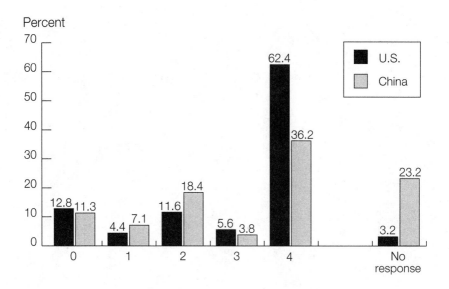